Collective Simulation-Based Training in the U.S. Army

User Interface Fidelity, Costs, and Training Effectiveness

Susan G. Straus, Matthew W. Lewis, Kathryn Connor, Rick Eden, Matthew E. Boyer, Timothy Marler, Christopher M. Carson, Geoffrey E. Grimm, Heather Smigowski

Prepared for the United States Army
Approved for public release; distribution unlimited

For more information on this publication, visit www.rand.org/t/RR2250

Library of Congress Cataloging-in-Publication Data is available for this publication.
ISBN: 978-1-9774-0132-8

Published by the RAND Corporation, Santa Monica, Calif.

Cover photos (clockwise from top left): U.S. Department of Defense photo; 1st Heavy Brigade Combat Team Public Affairs Office photo by PFC Aaron Braddy; U.S. Army photo; U.S. Army photo by Mr. Markeith Horace (Benning)

Preface

This report documents research and analysis conducted as part of a project entitled *Using Virtual Training Capabilities to Enhance Collective Training*, sponsored by the U.S. Army Training and Doctrine Command. The purpose of this study was to identify the levels of user simulator interface fidelity required to meet learning demands for crew and unit collective training and to provide systemic cost estimates of alternative approaches to providing levels of fidelity.

The Project Unique Identification Code (PUIC) for the project that produced this document is RAN167261.

This research was conducted within RAND Arroyo Center's Personnel, Training, and Health Program. RAND Arroyo Center, part of the RAND Corporation, is a federally funded research and development center (FFRDC) sponsored by the United States Army.

RAND operates under a "Federal-Wide Assurance" (FWA00003425) and complies with the *Code of Federal Regulations for the Protection of Human Subjects Under United States Law* (45 CFR 46), also known as "the Common Rule," as well as with the implementation guidance set forth in DoD Instruction 3216.02. As applicable, this compliance includes reviews and approvals by RAND's Institutional Review Board (the Human Subjects Protection Committee) and by the U.S. Army. The views of sources utilized in this study are solely their own and do not represent the official policy or position of DoD or the U.S. Government.

Contents

Figures

Tables

Summary

Motivation, Objective, and Approach

This report documents the results of a study sponsored by the U.S. Army Training and Doctrine Command, *Using Virtual Training Capabilities to Enhance Collective Training*. In the U.S. Army, the term *collective training* refers to training events in which groups or units of soldiers (e.g., teams, squads, platoons, companies, and above) learn, practice, and demonstrate proficiency in group activities key to their missions.

The Army perennially faces the challenge of efficiently and effectively conducting collective training. Virtual collective trainers are designed to help the Army meet this challenge and are the focus of our analyses and this report. Virtual equipment for simulation-based training (SBT) may consist either of (a) dedicated physical devices that resemble actual equipment, which we refer to as *physically simulated military equipment* (PSME), or (b) gamelike simulations that operate on personal computers, such as those available in the Army's Games for Training (GFT) program, which we refer to as *virtual military equipment* (VME) (Gorski and Parrish, 2017). For several decades, the Army has relied primarily on PSME for collective training. More recently, simulations involving games running on networked personal computers have evolved to the point that they rival or exceed the capabilities of legacy PSME simulators in many aspects of visual and aural realism and in authoring and feedback capabilities. VME approaches have less physical realism (an aspect of "fidelity") and may not fully replicate all modalities (e.g., touch, proprioception, and smell), but they are less expensive to operate, maintain, and upgrade.

Given these trade-offs between PMSE and VME approaches, the Army faces the question of whether it should change the way it conducts virtual collective training to take advantage of the potential savings VME offers. One potential area of transition for collective training of platoon- and company-level tasks is from a PSME-based approach—the Army's Close Combat Tactical Trainer (CCTT)[1] and the Aviation Combined Arms Tactical Trainer (AVCATT)—to a VME-based approach, currently Virtual Battlespace (VBS3)—the Army's flagship simulation GFT system. A transition to VME would likely significantly reduce the funding required to field, use, and maintain collective training capabilities. However, the decision to transition depends on understanding not only the cost of VME but its effectiveness relative to PSME. The implications of giving up the relatively high degree of physical fidelity that CCTT and AVCATT offer are of particular concern to the Army training community.

The primary objective of this research was to provide the Army with an improved understanding of simulator fidelity and its effects on learning platoon- and company-level collective skills. In addition to addressing the value of fidelity, we estimated the costs of collective training when using simulators with different degrees of fidelity, i.e., CCTT, AVCATT, and VBS3. *Simulation fidelity* is generally defined as the degree to which a simulator or simulation replicates real-world tasks, equipment, and environments. *Physical fidelity* reflects the degree to which simulators capture the look and feel of a system or environment in terms of such factors as equipment size and weight, appearance of controls, and cues or feedback (e.g., visual, auditory, and tactile). *Functional fidelity* reflects how the simulator behaves, such as the degree to which it responds to the user's actions in realistic ways or replicates functions of the actual equipment. *Psychological fidelity* refers to the degree to which the simulator or SBT prompts cognitive, behavioral, and affective responses relevant to performance in a particular setting, such as whether the training elicits the same need for attentional resources, perceptions of workload, and reactions or emo-

[1] The CCTT program includes a mobile variant called MCCTT. We use *CCTT* to refer to both systems and *fixed CCTT* when we specifically exclude mobile sets.

tions that occur when operating the actual equipment in a live environment.[2] These types of fidelity are not independent; for example, high levels of physical and functional fidelity can foster psychological fidelity (but are not required to do so). For the purposes of this report, *physical fidelity* comprises physical and functional dimensions because both are technology centered—determined by system hardware and software—while psychological fidelity is determined, in large part, by training strategy and design.

We conducted several analyses to evaluate collective SBT system fidelity and costs. To gain an understanding of system alternatives for collective training, we documented the use of SBT by the U.S. and British armies, including key attributes of the systems. To acquire a general understanding of the value of training system fidelity, we reviewed the research literature on how fidelity affects learning outcomes and other aspects of training effectiveness in military and commercial domains. To understand the views of key stakeholders—including senior managers of SBT systems; providers of air, ground, and dismounted training; and end users—we conducted interviews, focus groups, and surveys to gather data about critical features of such systems, how they are used, and their perceived strengths and weaknesses. To estimate the cost and the cost-effectiveness of the three training systems, we analyzed cost and usage data from key stakeholders and Army documentation.

Review of Prior Simulation Research

Figure S.1 summarizes the findings of our literature review. The figure shows that high psychological fidelity is key to effective learning and transfer of training. What matters most for designing effective SBT is fostering psychological fidelity by applying established learning principles. Findings for physical fidelity are complicated by the fact that many users perceive it to be important even though it may not contribute to training effectiveness. Research findings are clear that high levels

[2] We use *SBT* to reflect the training experience more broadly; the term encompasses not only hardware and software but also curricula and training support.

Figure S.1
Relationship Between Fidelity and Learning: Results of Research Review

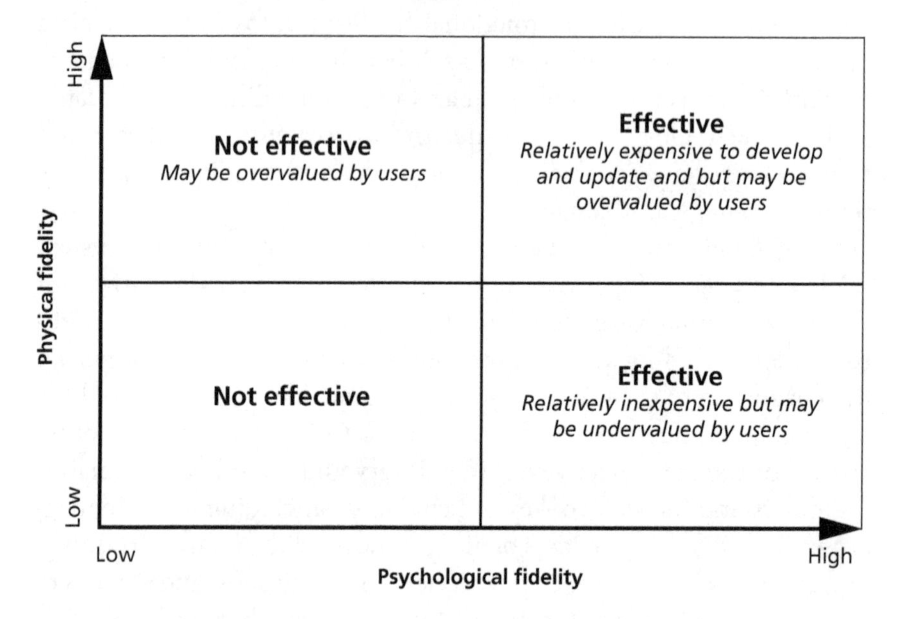

of physical fidelity do not necessarily produce better learning or transfer of training. Furthermore, simulators with high levels of physical fidelity impose substantial costs and may limit opportunities for practice, further diminishing their potential effectiveness. In contrast, lower physical fidelity, personal computer–based simulations are likely more cost-effective and can be more accessible,[3] thereby providing more opportunities for practice. Moreover, SBT can be effective with low physical fidelity devices if high psychological fidelity is achieved. This is especially the case for collective tasks, which are largely cognitive, involving such behaviors as monitoring others' actions, providing backup when others are overloaded, and synchronizing actions; cognitive skills decay more rapidly than psychomotor skills do and therefore require more fre-

---

[3] VBS3 is a government-off-the-shelf system based on a commercial-off-the-shelf game. The assumption of increased accessibility for personal computer–based simulations is based on the ease of use and reliability of commercial-off-the-shelf hardware and networks and on the development of software tools to support multiuser gaming, including authoring and feedback tools.

quent practice. Thus, the lower right quadrant of the figure represents a sweet spot for which training system developers should aim to capitalize on the effectiveness of creating high psychological fidelity while avoiding the expense of high physical fidelity trainers.

Although there have been advances in research on learning principles relevant to SBT, evaluation of training outcomes has focused on the performance of the simulator and relies on users' reactions to training or systems, rather than on the performance of trainees and objective measures of performance in training (Salas, Bowers, and Rhodenizer, 1998). This imbalance between human-centered and machine-centered concerns has persisted, with training technologies being developed much more rapidly than pedagogy (Cannon-Bowers and Bowers, 2009; Government Accountability Office, 2016; Stewart, Johnson, and Howse, 2008). This may, in turn, perpetuate assumptions about the importance of physical fidelity in SBT.

Stakeholder Views of Collective Simulation-Based Training Systems

Through the design, delivery, performance, and assessment of training, stakeholders develop potentially important insights regarding the relative values, strengths, weaknesses, and risks of training systems and regarding possible changes to the delivery of collective armor and aviation training. We sought such insights through data gathered during interviews, focus groups, and surveys with individuals representing several key stakeholder groups for Army SBT, including capability providers, training providers, and training consumers.

Respondents in interviews and focus groups revealed that Army stakeholder groups (developers, trainers, and consumers) held mixed views of both PSME and VME approaches. Representatives of each group reported that CCTT and AVCATT are valuable and important for platoon- and company-level collective training. Likewise, respondents to a survey administered to AVCATT users reported that AVCATT is valuable for collective training and that physical fidelity is important in SBT for collective tasks. However, this view was tempered by complaints that the physical fidelity of such systems lagged behind changes in actual military equipment platforms and by a reluctance to train on PSME devices that do not match standard operating equipment. Capability and training providers also raised concerns about the costs required to maintain the concurrency of AVCATT and CCTT with fielded equipment. Attitudes toward VME approaches were favorable among capability providers but were more mixed among training providers and consumers. Key drivers of utilization of all collective SBT systems include command emphasis, policy or requirements for use, commander knowledge and skill in using the systems and related training resources, and accessibility or ease of use. These factors are integral to the recommendations we make for future use of collective SBT systems.

Comparing the Usage Rates, Costs, and Cost Effectiveness of Collective Simulation-Based Training Systems

We compared the costs associated with the Army's existing systems for collective virtual training: CCTT, AVCATT, and GFT using VBS3. We also estimated and compared the cost-effectiveness of the training systems, although this analysis relies on utilization data and is limited, in part, by ongoing developments in the GFT program and by inconsistent reporting for GFT. That is, CCTT and AVCATT are long-established programs that are not adding training capacity. These programs have established utilization metrics and contractual requirements for reporting utilization. In contrast, GFT is a relatively new program and will not have stable capacity for several more years. And, unlike CCTT and AVCATT, GFT does not have centralized contractor support or contractual requirements for reporting utilization. In addition, our estimates of cost-effectiveness are limited by a lack of outcome data, i.e., data on quality of trainee performance in training.

Table S.1 shows the Army's current annual spending for CCTT, AVCATT, and the GFT programs. CCTT costs are much higher than GFT expenditures ($65.4 million and $36.3 million, respectively), even though the CCTT program is no longer fielding additional systems.

Table S.1
Games for Training Is the Least Expensive System, Particularly After Accounting for Utilization

	CCTT[a]	AVCATT[b]	GFT[c]
Total cost ($M)	65.40	52.30	36.30
Research and development ($M)	0.63	5.08	1.00
Procurement ($M)	43.97	34.65	6.35
Maintenance ($M)	3.47	0.02	0.20
Contract logistics support ($M)	17.30	12.51	—
Personnel ($M)	—	—	28.74
Cost per soldier potential training day at current utilization ($ per soldier day)	750.00	7,000.00	200.00

NOTE: All costs adjusted to fiscal year (FY) 2017 dollars.

[a] Average budgeted for FYs 2016–2021. Average actual expenditures for FYs 2010–2015 were lower, at $64.6 million, because of maintenance spending.

[b] Average budgeted for FYs 2016–2021. Average actuals for FYs 2010–2015 were lower, at $39.1 million.

[c] RAND Arroyo Center estimates for supporting personnel combined with GFT budget data.

Total AVCATT costs are higher annually than for GFT ($52.3 million and $36.3 million, respectively). AVCATT has high marginal costs for training, but these are far lower than the alternative of live training. However, GFT is also less expensive than CCTT and AVCATT when accounting for utilization (see "Cost per soldier potential training day at current utilization" in the table).

While the Army does not systematically collect training effectiveness data associated with each system, we developed measures that suggest their cost-effectiveness by combining available cost and utilization data. Such measures compare the systems in terms of cost per soldier day, as shown in Table S.1, as well as cost per task, event, and platoon hour. These measures indicate that CCTT and AVCATT are much less cost-effective than GFT. Moreover, GFT has applicability to various forms of collective training beyond what AVCATT and CCTT can offer.

Recommendations

PSME simulators are costly largely because they provide high levels of physical fidelity, and this cost may not be justifiable in terms of Army utilization rates, constraints on access to these systems, and extant research findings about the effectiveness of physical fidelity for training outcomes. However, definitive studies have not been conducted comparing CCTT and AVCATT with VME approaches on learning from training and training transfer for collective tasks.

Given these findings, we offer two broad, complementary types of recommendations. The first type pertains to improving SBT use, delivery, and evaluation regardless of the equipment used. The second type concerns the use of experiments to determine whether empirical evidence supports a transition from PSME systems (AVCATT and CCTT) to VME systems, such as VBS3, for collective training; if the evidence does provide support, the Army should begin this transition. We offer the following specific recommendations:

1. **Revise training policy and strategy to require or encourage use of collective SBT and supporting training support packages (TSPs), and begin to transition from PSME to VME.** Utilization of virtual training devices is higher with command emphasis and when training strategies prescribe their use. Command emphasis and changes to policy or strategy can also change the organizational culture, which is needed in light of persistent attitudes about the value of high physical fidelity in SBT for collective training.
2. **Improve and standardize measures of trainee performance in SBT.** Integrate objective measures that SBT systems collect automatically and the subjective measures that observer controllers collect using scorecards or other templates.
3. **Ensure that company leadership has access to comprehensive, high-quality TSPs, and provide institutional training and training support in units on the use of these TSPs.** TSPs should include scenarios and performance measures based on established educational principles and should leverage the poten-

tial of automated, objective feedback capabilities of simulators. Training support in units can include specific expertise similar to that of a tank master gunner, web-based train-the-trainer modules, and online user forums.

4. **Improve and expand the collection of utilization data to evaluate effectiveness of SBT at the program level, and support research on the optimal mix of collective training modalities.** Proposed data include standardized metrics of system utilization, costs, and training outcomes (learning and training transfer).
5. **Conduct one or more experiments or demonstration projects to test the relative effectiveness of PSME (AVCATT and CCTT) and VME (currently VBS3), along with robust TSPs, for collective training.** In addition to providing evidence about needed fidelity for training collective tasks, experiments or demonstration projects can garner end-user support and gather information to inform policy and support a transition to VME for collective training.
6. **After implementing recommendation 5, evaluate courses of action for continued use of CCTT, AVCATT, and VME approaches for collective training.** If, in contrast with what relevant literature suggests, there is evidence of greater training effectiveness for SBT using PSME rather than VME, the Army should evaluate the trade-offs of using the more-costly technologies, e.g., whether the additional training value of PSME justifies the higher annual costs. Evidence that effectiveness of SBT using VME is equal or greater to that using PSME would support phasing out CCTT and AVCATT. This process could begin at sites with low utilization by consolidating systems and placing them in warm storage while offering or increasing GFT capabilities by providing hardware and personnel. The transition should be supported with continued development of high-quality TSPs for all SBT, efforts to leverage objective performance feedback capabilities of VME, education and training on the value of VME and how to use the systems for collective training, and access to training support. Depending on the length of

the transition process, resourcing for CCTT or AVCATT at sites that still have access to these simulators should be aligned with anticipated utilization rates.

These recommendations are complementary in the sense that they become synergistic if the Army creates mechanisms to link them. For example, increased use of VME could be the source of information for prioritizing needed improvements to SBT—if the Army takes steps to ensure that VME utilization is appropriately measured and analyzed. Similarly, as SBT improves, utilization may increase—if the Army takes steps to ensure that commanders are made aware of and convinced of the improvements.

Acknowledgments

This projected benefited greatly from the support and guidance of COL Craig Unrath; the project's action officer, COL Scott Gilman; and members of the U.S. Army Training and Doctrine Command, Training Capability Manager Virtual & Gaming organization: LTC Michael Pavek, LTC Michael Stinchfield, Daniel Wakeman, Ruben Irizarry, MAJ Anthony T. Findlay, LTC Mike Martin, Daniel Finch, and MAJ Chris Finnigan. Colonel Gilman provided thoughtful stewardship of our efforts and access to many important resources and contacts that greatly added to the quality and timeliness of our research products. We also would like to acknowledge COL Jay P. Bullock, who supported this project following a change in command.

Our shared understanding of the history and literatures supporting virtual collective training was guided and strengthened by Eduardo Salas of Rice University and Joan Johnston and Paula Durlach of Army Research Laboratory. In their role as reviewers, Peter Schirmer of the RAND Corporation and Stephen Fiore of the University of Central Florida provided many valuable comments that improved the quality of this report.

We also thank the leadership and subject-matter experts at the Maneuver Center of Excellence and the Aviation Center of Excellence for hosting our team on visits; sharing their deep expertise; and providing access to background documents, training providers, and soldiers for interviews. Wade Becnel, of the Aviation Center of Excellence, and Paul Kizinkiewicz of the Maneuver Center of Excellence, were generous with their time and resources to support our efforts. Steve Krivitsky was instrumental in providing insights on the Integrated

Weapons Training Strategy and the formal integration of simulation-based training into future Army training.

Lee Harrison, of the U.S. Army's Program Executive Office for Simulation, Training and Instrumentation, provided support to the work at a number of points, and we benefited from his input during an observation of virtual collective training.

We are also indebted to members of the British Army's Training Branch for insights on their use of simulation-based training for ground forces. Col "Jonny" Ormerod, Lt Col Gwilym Vaughan, and Maj S. Roberts Anglian answered many questions and provided valuable background information on their training systems and outcomes.

Peter Khooshabeh, Julia Campbell, and CPT Andrew Jenkins of the University of Southern California's Institute for Creative Technologies were generous with their time and expertise to support our research.

Our data analyses of usage could not have been completed without the support of Christopher Holmes and Theodore Sellers of the Combined Arms Center–Training, Training Support Analysis and Integration Division, U.S. Army Training Support Center, who answered many questions and provided access to data.

We owe thanks to Dylan Sutro for consulting on the evaluation of computer game Arma 3, its mission editors, and after-action review module.

We offer a special thanks to the many soldiers and contractors who took the time to provide their opinions, concerns, and insights (without attribution) regarding their experiences with virtual collective training during our site visits. It is through their willingness to respond to our questions and speak frankly, whether face-to-face or via survey, that we were able to better understand the strengths, weaknesses, and opportunities for improvement in virtual collective training.

Many colleagues at RAND also offered useful feedback. 1st Lt Andrew Cady, U.S. Air Force, a Pardee RAND Graduate School doctoral student, generously shared his expertise from a RAND Project AIR FORCE project and provided access to content on U.S. Air Force simulation-based training. MAJ Stacy Moore, a fiscal year 2017 RAND Arroyo Center Fellow, provided contacts and support for our project. The work was strengthened by input from Bryan Hallmark and Michael Hansen at different phases of the project.

Finally, we would like to thank Donna White for her able assistance in all phases of this research and to Abby Schendt and Laura Coley for their expertise and efforts in producing this document.

CHAPTER ONE

Introduction

In the U.S. Army, the term *collective training* refers to training events in which units of soldiers (e.g., teams, squads, platoons, companies, and above) learn, practice, and demonstrate how to perform group activities that are key to their missions. Because it builds on individual training and prepares soldiers and smaller units for success in larger unit training events, collective training is central to the Army training cycle and is critical to the development and maintenance of the Army's readiness to conduct operations.

The Army perennially faces the challenge of conducting collective training efficiently and effectively. Currently, the Army uses four training modalities: live, virtual, constructive, and gaming (Army Regulation [AR] 350-1, 2014). This report focuses on virtual training and gaming.

Virtual training is characterized by real people operating simulated systems to perform such tasks as learning or exercising motor skills (e.g., flying a jet or driving a tank), making decisions, (e.g., when or where to execute fires), or communicating (e.g., among members of a command team).[1] For several decades, the Army has relied on virtual

---

[1] *Live* refers to simulations involving "real people operating real systems" (DoD, 2011, p. 119). Live training is considered to be a simulation because it is not conducted against a live enemy and may use some simulated equipment, for example, using lasers to simulate rounds rather than live ammunition. *Constructive* refers to

> simulated people operating simulated systems. Real people stimulate (make inputs to) such simulations but are not involved in determining the outcomes, for example, a user may input data instructing a unit to move and to engage an enemy target. The constructive simulation determines the speed of movement, the effect of the engagement with the enemy, and any battle damage that may occur. (DoD, 2011, p. 85)

systems—specifically, the Close Combat Tactical Trainer (CCTT)[2] and the Aviation Combined Arms Tactical Trainer (AVCATT)—to address virtual collective training requirements and supplement live, collective training for armor and aviation units, respectively. We refer to these virtual training systems as *physically simulated military equipment* (PSME), based on the Military Equipment Framework of Gorski and Parrish (2017).[3] Personal computer (PC)–based games have also evolved to provide highly realistic simulations. We refer to these training technologies as *virtual military equipment* (VME) (Gorski and Parrish, 2017). The Army's Combined Arms Center for Training's (CAC-T) Games for Training (GFT) program is developing VME via hardware and software platforms (currently using Virtual Battlespace 3 [VBS3]) and sets of networked laptop computers with headsets for communication. Compared to dedicated PSME, VME has less physical realism (an aspect of "fidelity") or immersion because soldiers do not need to enter a physical mock-up of a combat system to use it, and the systems may not fully replicate all modalities (touch, proprioception, smell, etc.). However, VME is less expensive to operate, maintain, and upgrade (Gorski and Parrish, 2017[4]).

Virtual training, constructive training, and gaming are considered to be supplements to, not replacements for, live training.

[2] The term *CCTT* can refer to a collective training program composed of three subsystems: CCTT, the Reconfigurable Vehicle Tactical Trainer (RVTT), and the Dismounted Soldier Training System (DSTS). In this report, we use the term to refer only to the core CCTT fixed and mobile units, not to the RVTT and DSTS (U.S. Army Acquisition Support Center, undated a).

[3] As we describe in Chapter Two, Gorski and Parrish (2017) also refer to what had been labeled "virtual and constructive" training as "synthetic training." In this report, we will continue to use *virtual* to refer to the collective training that is at the heart of our research, but we support future use of the terminology *synthetic environment using PSME* rather than *synthetic environment using VME*. We also discuss development of hybrids using both PSME and VME throughout this report.

[4] These authors acknowledge that training armor and aviation platoons and companies to make life-and-death decisions and to stimulate reactions that soldiers experience in combat is different from training in many other domains. Such training is most realistic in live training events, including live fire, as the closest proxy for the challenges of operating in combat.

As a result of these developments in technology for simulation-based training (SBT), the Army faces the question of whether it should increase the use of simulation approaches that use VME, rather than PSME, to take advantage of the potential savings a VME-based approach offers. One potential area of transition for collective training of platoon- and company-level tasks is to replace CCTT and AVCATT with VME-based training. This transition would likely significantly reduce the funding required to field, use, and maintain training capabilities. However, the decision to transition to VME depends on understanding not only its cost but also its effectiveness relative to CCTT and AVCATT. The implications of giving up the relatively high degree of physical fidelity that CCTT and AVCATT currently offer are of particular concern to the Army training community.

Understanding the effects of a shift from CCTT and AVCATT to a VME approach is also important as the Army transitions to the Synthetic Training Environment (STE).[5] The STE promises to provide capabilities beyond current virtual and constructive technologies to better represent operational complexity, increase accessibility, and lower costs. Among its goals are being less reliant on physical facilities and devices and being more transportable to deliver "training at the point of need." Based on Gorski and Parris's (2017) Military Equipment Framework, CCTT, AVCATT, and the GFT approach are all examples of providing training in a synthetic environment, but they differ in the types of equipment provided to the trainees: CCTT and AVCATT provide PSME, and the GFT approach, currently epitomized by VBS3, provides VME.

Objective and Approach

The primary objective of this research was to provide the Army with an improved understanding of the fidelity of SBT technologies and their effects on learning platoon- and company-level collective skills. Under-

[5] As of this report's publication, the STE program is in the pre–materiel development decision phase (PEO STRI, undated).

standing how fidelity contributes to training outcomes can help inform decisions regarding current and legacy systems and future investments in SBT for platoon- and company-level collective skills. The definition of *SBT* will be of central importance to our findings and recommendations: It includes the hardware, software, and development of associated curricula and training support. In addition to addressing the value of fidelity, we estimated the costs of collective training when using simulators with different degrees of fidelity. This report focuses on virtual collective trainers in use by U.S. Army armor and aviation units: CCTT, AVCATT, and GFT approaches using VBS3 for platoon- and company-level tasks.

Collective Training

CCTT and AVCATT were designed for collective training, which rides on the assumption that individual team members and individual teams are proficient in their technical areas of expertise, for example, as gunners, drivers, loaders, or commanders in tank crews or as pilots, flight engineers, or gunners in flight crews (Stewart et al., 2008; Swezey et al., 1998). Therefore, our focus in this report is on collective training at the platoon and company levels. We acknowledge that these training systems provide potentially valuable individual, team (subsets of crews), and crew training; however, other SBT systems focus on team-level tasks for crew members (e.g., tank commander and gunner or pilot and copilot or gunner).

An example of a representative collective task within this scope is "company movement to contact," in which tasks are distributed across the tanks and platoons that make up the company unit. The Combined Arms Training Strategy (CATS) for this task has subtasks listed in the performance measure column of Table 1.1. For example, to execute Subtask 9 of the "Movement to Contact" collective task requires the movement of a company of tanks in a formation. Examples of formations include line, column, staggered column, "vee," left flank, and right flank, herringbone, wedge, and modified wedge. Figure 1.1 shows a company-level modified wedge formation. Executing effective maneuver of a company-level wedge formation over terrain with potential obstacles and threats is a complex task involving collabora-

Table 1.1
Performance Steps for the "Conduct Movement to Contact (Platoon–Company)" Task

Performance Measure	Go	No-Go
1. Unit leader gained and/or maintained situational understanding.		
2. Unit leader received an operation order or fragmentary order, begins troop leading procedure and issues warning order to include at a minimum		
3. Unit leader prepared for movement to contact.		
4. Unit leader confirmed friendly and enemy situations.		
5. Unit leader performed the fundamentals.		
6. Unit leader issued clear and concise orders.		
7. The unit conducted a rehearsal.		
8. Unit leaders issued fragmentary orders that addressed changes to the plan.		
9. Unit executed a search-and-attack or cordon and search for one or more specified purposes.		
10. Unit leaders synchronized element actions.		
11. The unit consolidated and reorganized as necessary.		
12. The unit continued operations as directed.		

SOURCE: Adapted from Combined Arms Training Strategy Task Number 07-2-1090, U.S. Army, Combined Arms Training Strategies (CATS) database, accessed via the Army's Digital Training Management System, undated a.

tion, establishing shared situational awareness, and many other distributed tasks to monitor terrain and communicate within and above the company level.

A specific example of an instance of tactical decisionmaking demonstrates the complexity and need to integrate recognition, cognitive, decisionmaking, communication, and coordination skills in real time. At the Army's National Training Center (NTC), an armor company commander could be maneuvering a company of 14 vehicles over terrain during a "movement to contact." His or her task is to move toward an objective but be prepared to encounter enemy forces. The 14 tanks are divided into three platoons of four tanks each, plus the commander's own tank and that of the executive officer. They might be moving in one of several formations, including the modified wedge formation pictured in Figure 1.1.

The point of complex tactical decisionmaking comes when the front platoon, with the company commander behind it, crests a wadi and sees an antivehicle ditch that was not reported to them and what appears to be a hastily implanted minefield to the right of the ditch. The commander must immediately read the terrain and these obstacles, assess the situation, and provide initial guidance to the platoons in the company. This guidance will be based on his or her understanding and assessment of a number of factors, such as the enemy's likely goal (e.g., stop them? turn them into a kill zone?), positions for enemy antiarmor teams or direct fires, and authenticity of the minefield (real or decoy?). It will also be based on reports from the platoons of any sightings of enemy positions or fires. The commander must then make a series of

Figure 1.1
Company of Tanks in a Modified Wedge Formation

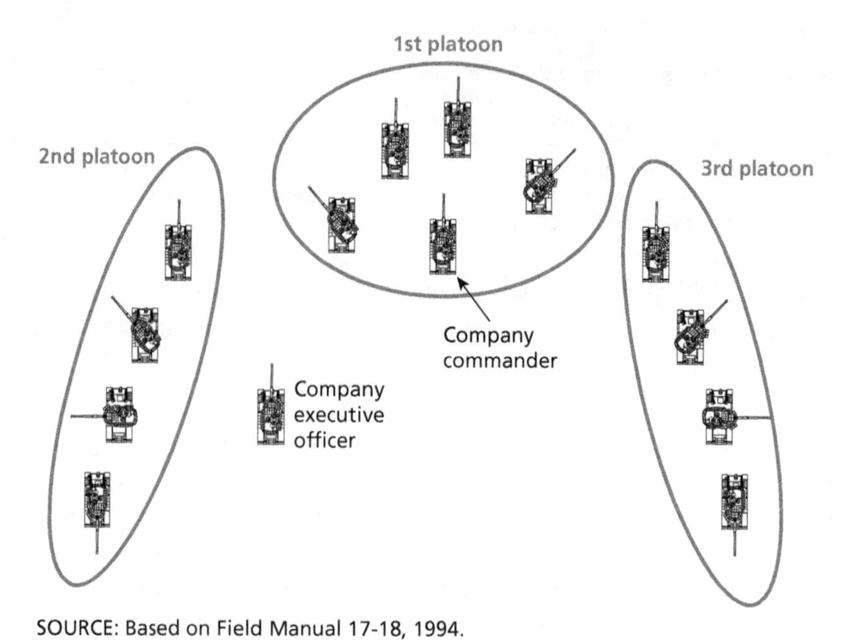

SOURCE: Based on Field Manual 17-18, 1994.

decisions (e.g., how to orient the tanks in each platoon; whether to withdraw, turn, or try to breach the obstacles) and communicate these decisions to the platoons, which in turn must execute these actions. All these perceptions, inferences about the enemy's intent, decisions, communications, and execution of actions must be made rapidly if the company is to achieve its mission.

Appendix A provides additional examples of collective tasks.

Simulation Fidelity

Simulation fidelity is generally defined as the degree to which a simulator or simulation replicates real-world tasks, equipment, and environments (e.g., Bowers and Jentsch, 2001). Scholars have identified and operationalized elements or types of fidelity in myriad ways (see Liu et al., 2008, for a review). Three types or dimensions of fidelity discussed frequently in the literature are physical, functional, and psychological fidelity (e.g., Bowers and Jentsch, 2001; Dietz et al., 2013).

Physical fidelity reflects the degree to which simulators capture the "look and feel" of a system or environment. Such factors as equipment size and weight, positions of actors (e.g., seat locations in a cockpit or tank simulator), appearance of controls, and cues or feedback (e.g., visual, auditory, olfactory, vestibular, tactile, haptic) are elements of physical fidelity. *Functional fidelity* reflects how the simulator "acts," such as the degree to which a simulator responds to the user's actions in realistic ways (e.g., Allen, Hays, and Buffardi, 1986) or replicates functions of the actual equipment (e.g., Bowers and Jentsch, 2001).[6] For

[6] *Functional fidelity* is defined in different ways in the literature. As described above, some scholars view functional fidelity as the realism with which the system responds to user actions and/or what the system allows users to do, for example, to operate equipment or perform functions (e.g., communicate) in the same way as one would using the actual equipment (e.g., Allen, Hays, and Buffardi, 1986; Bowers and Jentsch, 2001). This view of functional fidelity is closely aligned with physical fidelity because it reflects features built into the system. In fact, we argue that many of the cues typically associated with physical fidelity also play a role in functional fidelity. Others view functional fidelity as the degree to which SBT replicates the "purpose, meaning, and other situational or contextual parameters surrounding the task" (Dietz et al., 2013, p. 353), in terms of such factors as mission goals and roles or the degree to which the user must think and behave as he or she would when using the actual equipment in a live environment (Salas, Paige, and Rosen, 2013). This view

example, an aircraft simulator that allows simulation of engine failure or gives the sense of climbing in altitude after the user pulls back on the yoke is an example of functional fidelity (Bowers and Jentsch, 2001; Liu et al. 2008). *Psychological fidelity* (which is sometimes referred to as *cognitive fidelity* or *psychological-cognitive fidelity*) refers to the degree to which the simulator or simulation prompts cognitive, behavioral, and affective (emotional) responses relevant to performance in a particular setting (Bowers and Jentsch, 2001; Kozlowski and DeShon, 2004). For example, SBT with high levels of psychological fidelity elicits needs for attentional resources, perceptions of workload, and reactions or emotions similar to those that occur when operating the actual equipment in a live environment.

These dimensions of fidelity are not independent. High levels of physical and functional fidelity can foster psychological fidelity but are not necessarily required to do so. Psychological fidelity is based on the purpose and application of SBT (Bowers and Jentsch, 2001); it is created, in large part, by training strategy and design (Kozlowski and DeShon, 2004), which we describe in Chapter Three.

In this report, we refer to *physical fidelity* as comprising physical and functional dimensions. Both physical and functional fidelity are technology-centered because they are determined by system hardware and software, and both pertain to the realism of the equipment. Likewise, other authors do not distinguish between physical and functional fidelity, e.g., Kozlowski and DeShon (2004), and we have found few studies of simulators that examine functional fidelity as distinct from physical fidelity.

Approach

We conducted four tasks to assess needed levels of fidelity in SBT and associated costs. First, to gain a current understanding of the three simulator alternatives, we documented the use of simulators by the U.S. and British armies, including key attributes of the systems and costs associated with system acquisition, use, and maintenance. Second, to

---

of functional fidelity is more closely aligned with psychological fidelity because it is elicited by training design and is not dependent on system hardware or software.

acquire a general understanding of the value of training system fidelity, we reviewed the research literature on how fidelity affects learning outcomes and other aspects of training effectiveness in military and commercial domains. Third, to understand the views of key stakeholders—including senior managers of SBT systems, providers of air, ground, and dismounted training, and end users—we conducted interviews, focus groups, and surveys to gather data about critical features of such systems, how they are used, their strengths and weaknesses, and areas for improvement. Fourth, to estimate cost and the cost-effectiveness of the alternatives, we gathered cost and usage data from key stakeholders and Army documentation.

Organization of This Report

Chapter Two describes the three SBT systems that we compare: CCTT, AVCATT, and VBS3. We describe key attributes of each system, goals for system use, and how the systems are used in practice, including usage statistics in the Army. In addition, we describe use of virtual collective training systems by the British ground forces.

Chapter Three presents the results of a review of the literature regarding how fidelity in SBT affects training outcomes.

Chapter Four describes how key stakeholders view the relative advantages of the three systems, including their views on different aspects of fidelity.

Chapter Five presents the results of our analyses of the relative costs and cost-effectiveness of CCTT, AVCATT, and VBS3.

Chapter Six summarizes our key findings, offers recommendations for collective training to enhance learning and reduce costs, and proposes directions for future research.

Appendix A presents examples of tasks taught in collective training, while Appendixes B and C present questions used in interviews and in focus groups of key stakeholders and surveys of end users, respectively, as reported in Chapter Four.

CHAPTER TWO

Current U.S. Army Collective Simulation-Based Training Systems

This chapter describes and compares CCTT, AVCATT, and VBS3 as systems for providing collective training for armor and aviation platoon- and company-level tasks. The first half of the chapter focuses on the physical characteristics of the systems as alternative technologies. We begin by placing the three systems within a military equipment framework (Gorski and Parrish, 2017) for defining and comparing different types of simulation fidelities. We then describe CCTT, AVCATT, and VBS3 and compare their key attributes. For comparison purposes, we briefly describe the British Army's system for providing collective SBT. We end this comparison by summarizing the Army's relative capacity of each system in terms of the number and distribution of training sets.

The Military Equipment Framework Distinguishes Training Environments and Types of Simulated Equipment

Gorski and Parrish (2017) defined a military equipment framework to create a common vocabulary to improve communication among stakeholders in SBT. Their framework is based on Milgram and Kisihino's (1994) "virtuality continuum." This framework does not make a distinction between virtual and constructive simulations but instead refers to training environments that are either live or synthetic.

Within the synthetic training environment, the framework distinguishes two types of equipment that can be present. *Simulated military equipment* includes physical equipment or interfaces, such as the manned modules in CCTT and AVCATT,[1] and the simulated subsystems in the Conduct-of-Fire Trainer; by contrast, *VME* includes digital simulations and visual representations of military equipment.[2]

In terms of Gorski and Parrish's framework, all three systems—CCTT, AVCATT, and VBS3—provide training in a synthetic environment. However, CCTT and AVCATT use PSME, and VBS3 uses VME via government-off-the shelf (GOTS) interfaces, based on a commercial-off-the-shelf (COTS) simulation-based armor combat game. In the future, however, hybrid systems that integrate PSME and VME will become more prevalent. We discuss hybrid systems in more detail in Chapter Four.

Three Collective Simulation-Based Training Systems

CCTT is the Army's simulation system that targets collective tasks for Army armor, mechanized infantry, cavalry, infantry, and reconnaissance crews, as well as two types of trucks. AVCATT is the Army's simulation system for aviation crews for a variety of airframes (U.S. Army Acquisition Support Center, undated b). VBS3 can be used to train collective skills for the same training audiences but also trains a broader range of tasks. In the following subsections, we discuss the key elements of each system, as well as the British Army's system for providing collective SBT, the United Kingdom Combined Arms Tactical Trainer (UK-CATT).

CCTT

CCTT is a collective training program with roots in the Army's original, networked battlefield simulator system called SIMNET (for "sim-

[1] As well as the simulated weapons in the Engagement Skills Trainer.

[2] Note that others have abbreviated "simulated military equipment" as "SME." To avoid confusion with "subject-matter expert (SME)" in this report and to clarify the focus on physical equipment, we instead use PSME.

ulator networking"), which was fielded beginning in 1987. SIMNET was used for training until its successor, CCTT, became operational in the mid-1990s (Johnson, Mastaglio, and Peterson, 1993). The U.S. Army's CATT family of combined arms collective training simulation systems originally included both CCTT and AVCATT (Headquarters, Department of the Army, 2015). A cost and training effectiveness analysis for CCTT documented that the costs to develop and field the armor training capabilities of CCTT were justified by trading live training tank "operational tempo miles" funding for the CCTT development (Noble and Johnson, 1991).

The Program Executive Office for Simulation, Training and Instrumentation's (PEO STRI's) CCTT program currently has three subprograms:

- **CCTT** consists of computer-driven, manned module simulators replicating the vehicles found in combat units, including the M1 Abrams Tank, the M2 Bradley Fighting Vehicle (BFV), and the M3 Cavalry Fighting Vehicle (PEO STRI, 2016). CCTT is a mature program that is not expanding sites.
- **RVTT** consists of vehicle and mounted weapon simulators to provide collective virtual training to units. Training audiences are crews through the platoon level that are training on tactics, techniques, and procedures in wheeled maneuver collective training tasks, including convoy operations and mounted patrols (PEO STRI, 2016).
- **DSTS** consists of a networked, squad-size set of virtual reality tracking and visual displays for individual soldiers (Bymer, 2012). The goal is to provide practice in squad-level collective tactics for dismounted soldiers.

The first subprogram, CCTT, consists of fixed and mobile facilities (unless otherwise indicated, we refer to both systems as CCTT in this report). The fixed facility is equipped as follows:

- 14 Abrams M1A2 System Advanced Package version 2 Common Remotely Operated Weapon Station modules

- 14 Bradley M2A3 Chassis Modernization/Embedded Diagnostics modules
- two Bradley Fire Support Team Vehicle M2A3 Chassis Modernization/Embedded Diagnostics modules with Fire Support Sensor System (as of fiscal year [FY] 2016)
- one High Mobility Multipurpose Wheeled Vehicle Simulator
- five technology-enabled after-action review (AAR) capabilities.

The mobile CCTT (MCCTT) equipment consists of five standard tractor-trailers that provide for four-tank or –infantry fighting vehicle platoon training for M1 and M2 crews. This equipment is used to provide virtual training to Army National Guard (ARNG) armor units that cannot train at a fixed facility. Our research included analysis of the performance and costs of CCTT and MCCTT, which are much more costly than RVTT and DSTS and are the primary focus of SBT for armor units.

The CCTT manned modules are housed in warehouse-sized facilities that also contain classrooms and training and AAR rooms. Figure 2.1 shows the exterior and interior of a CCTT manned module.

It is important to note that, even though CCTT and AVCATT have high physical fidelity, they are not perfect replicas of the actual Army equipment. For example, in actual M1 tanks, the loader can stand in the turret during maneuver, scan for threats to the left and rear, and operate an external machine gun. CCTT's M1 tank simulators do not allow the loader to rise out of the turret and provide security and visual overwatch to key parts of the tank's battlespace, and the loader does not handle simulated rounds. Instead, the loader selects the specified round type by pressing the appropriate button in the ammunition storage area and then hits a switch on the main gun's simulated breech. In addition, as we describe in Chapter Four, the visual displays of the "popped hatch" are reportedly of insufficient quality for tank commanders to fight their tank as they would at the NTC, i.e., from standing in the turret and having their torso outside the open hatch.

Figure 2.1
Exterior and Interior of Close Combat Tactical Trainer Manned Crew Modules

SOURCE: U.S. Army Training and Doctrine Command (TRADOC) Capability Manager (TCM) Virtual and Gaming (V&G) Operations. The left image is a RAND photograph.
NOTE: On the left is a room containing the networked CCTT and manned crew modules. On the right is an interior view of a manned crew module.

AVCATT

AVCATT is the Army's collective training system designed to meet aviation training requirements and support institutional, organizational, and sustainment training and mission rehearsal. As with CCTT, AVCATT is a mature program that is not expanding to new locations.

As mentioned earlier, AVCATT was originally part of the CATT family of SBT systems, along with CCTT. As with MCCTT, AVCATT is mobile via trailers that house networked simulators. However, unlike MCCTT's M1 and M2 mobile training modules, AVCATT modules are "reconfigurable" to provide training on attack, reconnaissance, lift, and cargo helicopters and to provide role-player stations for battalion and squadron staff or others (U.S. Army Acquisition Support Center, undated b). The mobile unit set also contains an AAR theater and a battle master control station.

As noted earlier for CCTT, AVCATT has high physical fidelity but does not perfectly replicate actual Army equipment. For example, the Apache AVCATT trainer has pilot and gunner sitting side by side and has a flat control panel; in an actual Apache helicopter, the crew sits tandem, and the control panels are shaped differently. Figure 2.2

presents a photograph of an actual Apache cockpit on the left and one of AVCATT on the right. Figure 2.3 presents a diagram of the two elements of the AVCATT trailer set.

VBS3

VBS3 is a system within the Army's GFT program, that seeks to provide "low overhead, easily adaptable and readily available, commercial

Figure 2.2
Actual and Simulated Apache Helicopter Cockpits

SOURCES: Karl Drage, photographer, Global Aviation Resource; used with permission (left). RAND photograph (right).
NOTES: Actual Apache Helicopter Pilot cockpit is on the left; the AVCATT physical simulator for that cockpit is on the right.

Figure 2.3
Diagram of Aviation Combined Arms Tactical Trainer Trailer

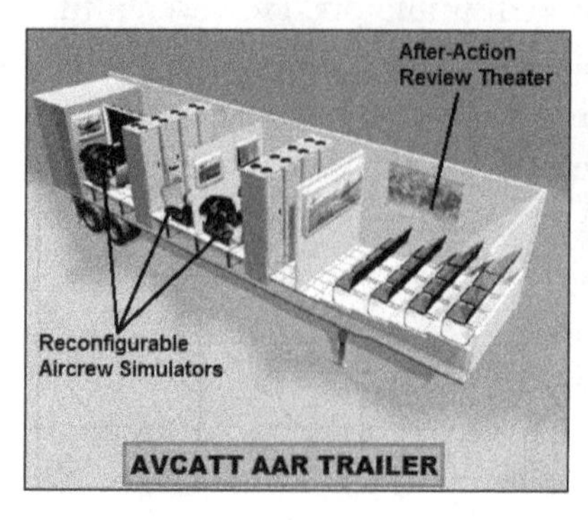

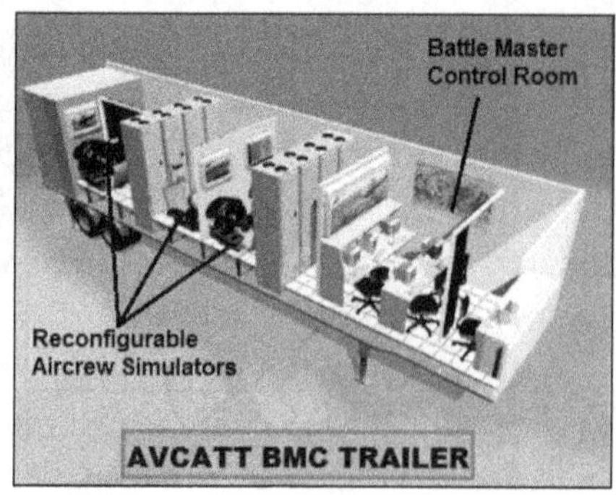

SOURCE: PEO STRI, 2016.
NOTE: BMC = battle master controller.

off-the-shelf (COTS) and GOTS gaming applications" (PEO STRI, 2016, p. 75) to support training and mission rehearsal.[3] In contrast to CCTT and AVCATT, which are PSME, VBS3 is a VME-based training system designed to support SBT of a wide range of individual and collective skills. To date, GFT has provided a suite of training software applications installed on PC-based, networked, multiplayer laptop and desktop computers. The GFT program has developed over 92 training support packages (TSPs) covering 103 collective tasks, accessible via a web portal. Figure 2.4 depicts a group of stations that are part of a GFT set of hardware.

VBS3 has been described as the "U.S. Army's flagship gaming engine."[4] The system is based on a popular COTS first-person-shooter

Figure 2.4
Four Games for Training Stations Deployed in a Room: Four Networked Gaming Laptops and Headsets with Microphones

SOURCE: RAND photo.

---

[3] In addition to VBS3, the GFT program includes trainers for bilateral negotiation (BiLAT), Tactical Iraqi language and culture training (Tactical Iraq), and tactical language skills (Pashto and Tactical Dari).

[4] U.S. Army, "Virtual Battlespace 3," Stand-To website, May 19, 2014.

game called Arma 3, created by Bohemia Interactive (Bohemia Interactive, undated). VBS3 is described as a three-dimensional

> first-person games-for-training platform that provides realistic semi-immersive environments; large, dynamic terrain areas; hundreds of simulated military and civilian entities; and a range of geotypical, or generic, terrain areas as well as geospecific terrains from U.S. Army areas of operation. As the Army's flagship training game, it has been accredited to support more than 100 combined arms training tasks from the individual soldier level to company collective. (Gourley, 2016)

VBS3 can be operated to carry out individual, crew, team, platoon, and company collective tasks by armor and aviation crews and to train skills for other proponents. For our comparisons, we will refer to VBS3 being used to execute collective tasks at the platoon and company levels. The visual resolution of VBS3 was significantly higher than that of the original CCTT image generator, so, beginning in 2015, the VBS3 Image Generator was fielded as an update to the CCTT M1 modules' image generator (Bohemia Interactive Simulations, 2014). The semiautomated forces for CCTT and AVCATT are controlled using the Army's OneSAF system.

Figure 2.5 is a VBS3 screen image from a simulation involving armor operations.

GFT approaches are most commonly accessed using COTS interface hardware (a commercial headset with headphones and a microphone for communications, keyboard, and mouse for interaction with the simulation) that, for security reasons, is not connected to a central server. Custom-developed PSME interfaces, GOTS interfaces to access VBS3, could be configured as hybrids to include the sighting periscope housing and gun control "Cadillac" yoke for the M2 gunner; separate tank commander controls, as pictured in Figure 2.6; and realistic tanker helmets and communication control switches. Such high-physical-fidelity interface peripherals are part of the current tabletop Conduct-of-Fire Trainer system for the M2 (see Figure 2.6), which is a PSME system trainer for the Bradley gunner and tank commander. Experiments with hybrid approaches are being carried out at the Com-

Figure 2.5
Virtual Battlespace 3 Display of Simulated Armor Operations

SOURCE: Virtual Battlespace 3 Display of Simulated Armor Operations © 2007–2018 Bohemia Interactive Simulations, k.s. All rights reserved.

Figure 2.6
Tabletop Conduct-of-Fire Trainer

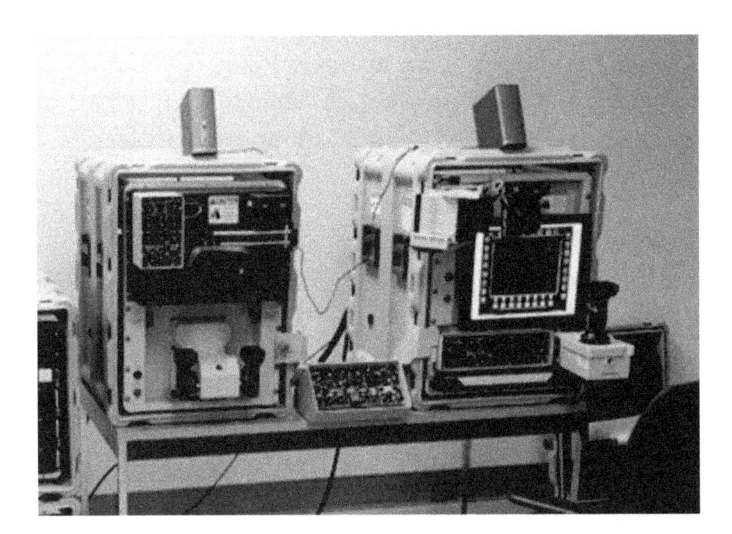

SOURCE: RAND photo.

bined Arms Center's Training Innovation Facility at Fort Leavenworth, Kansas.

One of the motivations for using a VME approach to deliver collective training was to address the costs of updating training equipment as changes are made to actual equipment. Updating the fleet of 260 PSME armored vehicle manned modules in CCTT, for example, is costly, as is making the requisite changes to the software to simulate the changed hardware. In contrast, VME software can be changed and rolled out in a single, systemwide update (although, because these systems are not connected to servers, this process is not automated and requires local intervention). Additionally, the GFT laptops can be physically oriented in roughly the same crew positions as in the actual equipment, e.g., the laptops for the pilot and copilot or gunner can be positioned on tables so that the pilot sits physically behind the copilot or gunner, as they would sit in tandem in the actual airframe.

UK-CATT

The British Army has employed what it refers to as virtual reality simulation for collective training since 2003, when it opened two UK-CATT sites, one at the UK Ground Forces Land Warfare Centre, Warminster, England, and the other in Sennelager, Germany (Ormerod, 2015).

UK-CATT is similar to the U.S. Army's CCTT but differs in a number of important ways. Chief among these differences are the following:

- The scale of training events that can be supported: UK-CATT provides training for brigade-level maneuver and mission training. This capability was acquired because of the limitations on physical training spaces in British Isles: They have no local NTC or Joint Readiness Training Center equivalents, except at training sites in Canada and Kenya.
- PSME: The platforms simulated are British Army combat vehicles and helicopters.

UK-CATT links 140 simulators and up to 450 participants to provide collective training to armor, mechanized infantry, dismounted infantry, and all echelons of command tactical operations centers up to the division level along with aviation assets (Nash, 2016). It also allows soldiers to interact with computer-generated civilians, as well as friendly and enemy forces.[5]

As with CCTT, AVCATT, and the GFT approach, findings from discussions with the British capability providers, training providers, and consumers of training are shared in Chapter Four.

Number and Distribution of CCTT and AVCATT Sets and GFT Suites

The Army's collective training systems, reviewed earlier, have the capacity to provide collective training at the platoon and company level on armor and aviation tasks simultaneously to different numbers of crews or soldiers. The capacities of these systems depend largely on the number and distribution of the training sets and suites that the Army invests in (see Table 2.1).

From this point forward in the report, we use the term *set* (rather than both *set* and *suite*) to facilitate comparison.

The number of locations for collective SBT and component alignments varies (in FY 2016, fixed CCTT had seven; MCCTT had seven; AVCATT had 22; and GFT had 99) and are located at both continental U.S. (CONUS) and outside CONUS (OCONUS) sites, with some locations having more equipment than others, as shown in Table 2.2.

How Army Virtual Collective Trainers Are Used: Skills Trained and Context

Table 2.3 lays out the different levels of training (individual, team or crew, platoon or company) and the different types of systems reviewed. Cells shaded in dark gray indicate the primary training level targeted by designers; cells shaded in light gray indicate where systems may have

---

[5] Note that, with respect to costs, UK-CATT reportedly has

> evidence to suggest that CATT is 10 times cheaper than conducting Live training in terms of running costs, although the initial set up costs were large. As an alternative, a desktop version of CATT (used by most of Eastern Europe) using VBS2/3 is considered to be 100 times cheaper than live training and initial set up costs might be more than 100 times less than CATT; however, the level of fidelity of the simulator is greatly reduced. (Ormerod, 2015, p. 2)

Table 2.1
Comparison of Training Equipment Sets

	CCTT Set	MCCTT Set	AVCATT Set	GFT Sets
Number of vehicles simulated (simultaneous)	Two companies: 28 vehicle modules (14 each M1 and M2/3) in most sets	One platoon: Four vehicle modules (M1 or M2/3)	Six reconfigurable helicopter simulators	Three 52-seat sets (up to 39 vehicles)
Maximum number of soldiers	98	16 (M1) or 12 (M2/3)	12	156[a]
Dismounts	No	No	No	Yes

SOURCE: CAC-T, 2014.

NOTES: The M1 has a crew of four, the M/2/3 has a crew of three. The helicopter simulators have a crew of two.

[a] Soldiers with options for vehicle type or dismount.

Table 2.2
Summary of Training Equipment, Number, Components, and Locations in FY 2016

		CONUS		OCONUS	
Simulator	Component	Sets (no.)	Locations (no.)	Sets (no.)	Locations (no.)
Fixed CCTT sets[a]	AC[b]	6	6	1	1
	ARNG	0	0	0	0
MCCTT sets[a]	AC	0	0	0	0
	ARNG	14	7	0	0
AVCATT[c]	AC	10	10	2	2
	ARNG	11	11	0	0
GFT sets[d]	AC	45	25	10	6
	ARNG	67	67	1	1

[a] Data supplied by TSAID U.S. Army Training Support Center. The fixed CCTT and MCCTT counts are from 2016; however, the numbers of sets and locations can vary by year.

[b] AC = active component.

[c] Data supplied by TSAID U.S. Army Training Support Center. Ten active-duty AVCATT sites were reported in the 2nd quarter FY 2016 utilization report. Only nine active-duty sites were reported in the 3rd quarter. Eleven ARNG AVCATT sites reported in both the 2nd and 3rd quarters.

[d] Data supplied by PEO STRI.

value for training but were not the primary targets of the design. That is, the designers of live training targeted their training at all levels, from individual to collective platoon or company training, and the designers of CCTT and AVCATT designed their training at the platoon or company level, shaded in dark gray. However, while these systems were not designed or accredited to meet the system training aids, devices, simulators, and simulations (TADSS) fidelity requirements for individual and crew tasks required prior to live gunnery, soldiers have found these systems to have some utility at the team or crew and individual levels, as indicated by the light gray shading. Similarly, the GFT designers

Table 2.3
Training Target Levels Vary by Emphasis of Simulation-Based Training System: Individual, Team or Crew, and Platoon or Company

	Type of Training		
Training Target	**Live**	**Virtual with PSME**	**Virtual with VME**
Individual	Live fire Laser engagement	CCTT AVCATT	GFT
Team or crew	Live fire Laser engagement	CCTT AVCATT	GFT
Platoon or company	Live fire Laser engagement	CCTT AVCATT	GFT

NOTE: Dark gray cells indicate the primary training level targeted by designers; light gray cells indicate where systems may have value for training but were not the primary targets of the design. *Live fire* refers to the use of real bullets or tank rounds against simulated (e.g., wooden) targets. *Laser engagement* refers to the use of lasers instead of live ammunition against humans or vehicles with humans onboard.

targeted platoon or company and team or crew training, but users have found some utility at individual level.

A 2014 CAC-T study comparing CCTT and GFT systems supports the use of both CCTT and GFT for training armor collective tasks. This study considered whether CCTT and GFT support the training and evaluation outline (T&EO) only for collective tasks. The findings from the study, reported in Table 2.4, suggest that a GFT approach can provide training on a larger number of tasks (48 of 50 tasks) than can CCTT (12 of 50 tasks). The study did not directly measure training outcomes or assess whether one system trained better than the other. However, for 50 supporting collective tasks in a sample of six armor companies engaging in armor brigade combat team (ABCT) situational training exercises, SMEs judged that GFT was more capable than CCTT of meeting criteria of 38 performance measures (see Figure 2.7).

Table 2.4
Tasks Trained and Quality of Training Rated by Subject-Matter Experts for Close Combat Tactical Trainer and Games for Training

Metric	CCTT	GFT
Number of tasks trained	12	48
SMEs rating technology as providing superior task performance	2	38

SOURCE: CAC-T, 2014.

Figure 2.7
Close Combat Tactical Trainer and Games for Training Task Comparison of Six Sampled Situational Training Exercises

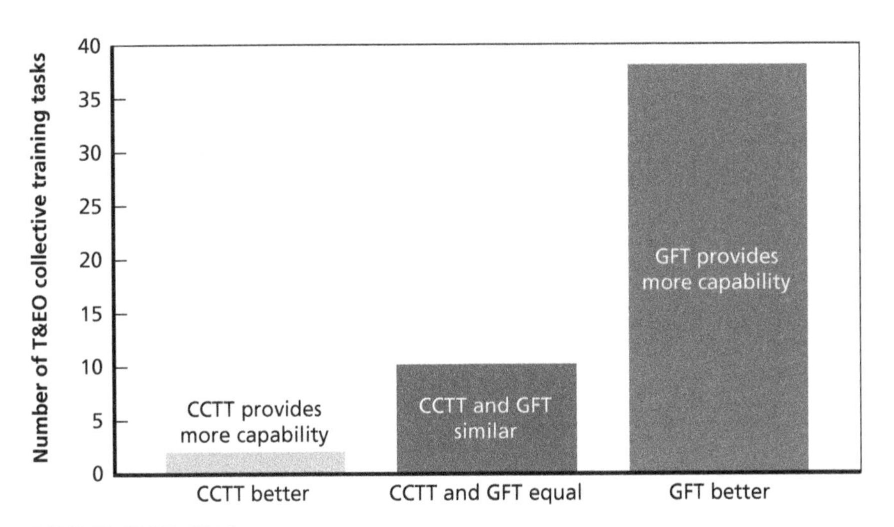

SOURCE: CAC-T, 2014.

Summary

The U.S. Army's training systems that our research focused on—CCTT, AVCATT, and the GFT-approach system, VBS3—provide synthetic training for collective tasks. The features of these systems suggest that VME provides greater capabilities for collective SBT. CCTT and AVCATT have physical fidelity but do not replicate all aspects of fielded systems, e.g., physical movement of the vehicle, maneuvering with the tank commander out of the hatch, and loader tasks. VBS3 lacks the physical fidelity features of CCTT and AVCATT but is flexible in terms of how the equipment is set up to mirror positions of crew members in physical space, and the software is easier to update to reflect user interfaces. In comparison with CCTT and AVCATT sets, GFT sets can easily scale to train a larger number of trainees simultaneously by adding equipment. Finally, CCTT and AVCATT were designed primarily to provide collective training at the platoon and company level, but VBS3 was designed to provide training at individual, team and crew, and platoon and company levels and can train more tasks.

We next provide findings from past research that inform the training of collective skills.

CHAPTER THREE

Research Findings on Simulation Fidelity and Training Outcomes

In this chapter, we review research on fidelity and learning outcomes for collective training. Whereas the past three decades have seen a substantial amount of research on SBT for teams, there is much less research on training for teams of teams (e.g., Mathieu, Marks, and Zaccaro, 2001). A *multiteam system* (MTS) consists of two or more tightly coupled teams that interact and are interdependent with respect to a common goal. Each component team might, at the same time, be pursuing its own goals. Thus, much of the research relevant to team performance on such topics as communication and coordination is relevant to the activities and behaviors that are critical for effective performance of an MTS. In fact, in an MTS with highly interdependent component teams, as in the company- and platoon-level collective tasks we focus on in this report, cross-team communication and coordination are especially important for system effectiveness (Marks et al., 2005).

This chapter is not intended to be a comprehensive review; instead, our goal is to present summative findings regarding fidelity and learning outcomes. Consequently, we focus largely on sources that presented meta-analyses or other reviews of relevant research. We reference discrete studies on topics for which quantitative or qualitative summaries are not available or to illustrate examples of particular findings. We limited the scope of these sources to studies involving real-world participants (e.g., aviators, emergency responders); with few exceptions, we did not include sources that used students as research participants unless the participants were students enrolled in professional training programs engaged in job-relevant training.

To identify sources for this chapter, we started with a corpus of sources that SMEs had recommended, then used a snowball approach, reviewing sources cited in the initial collection of documents, then those cited in subsequent documents.

As discussed in Chapter One, we use *physical fidelity* to refer to physical and functional capabilities of simulators and *psychological fidelity* to refer to the capabilities of the simulator or simulation to elicit the cognitive, behavioral, and affective responses relevant to behavior in actual task environments.

Training outcomes can be measured in different ways (e.g., Alvarez, Salas, and Garafano, 2004; Kirkpatrick, 1994; Kraiger, Ford, and Salas, 1993). Training evaluations are often limited to assessing user reactions using end-of-course surveys. Affective reactions (how much trainees report enjoying a course) do not predict learning (Alliger et al., 1997) but might influence motivation to participate in subsequent training (Bell et al., 2017).[1] In contrast, although reactions in the form of trainees' attitudes toward the utility of training (beliefs that the training will affect trainees' ability to do their jobs)) are related (albeit modestly) to learning and training transfer (i.e., application of learned knowledge and skills to the job) (Alliger et al., 1997), extant research shows that beliefs about the value of physical fidelity are inconsistent with findings based on more-objective measures of training outcomes. Therefore, in assessing the impact of fidelity on training outcomes, we focus on learning and behavior in terms of acquisition of knowledge and skills from training, knowledge and skill retention, and transfer of training.

Figure 3.1 summarizes the findings of our review. As the figure shows, high psychological fidelity is key to effective learning and transfer of training. What matters most for designing effective SBT is fostering psychological fidelity through the application of established learning principles. The findings for physical fidelity are complicated by the fact that users may perceive that it is important even though

[1] One exception is Warr, Allan, and Birdi (1999), who found that affective reactions to training predicted a change in immediate learning outcomes but not in behavior.

Figure 3.1
Relationship Between Fidelity and Learning: Results of Research Review

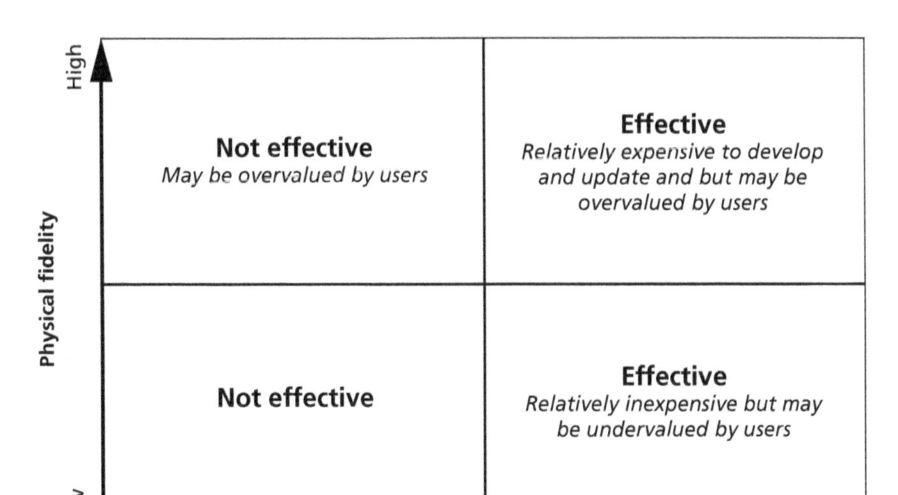

NOTE: While the combination of high physical and high psychological fidelity in SBT is generally effective, there are some caveats to this conclusion; for example, high physical fidelity can be detrimental to learning for novices but not for experts (e.g., Hays and Singer, 1989; Noble, 2002).

it may not contribute to training effectiveness in terms of changes in learner knowledge or behavior or training transfer. Research findings are clear that high levels of physical fidelity do not necessarily produce better learning or transfer of training. Furthermore, simulators with high levels of physical fidelity impose substantial development, operating, and maintaining costs and may limit opportunities for practice, further diminishing their potential effectiveness. In contrast, lower physical fidelity simulations, such as VME approaches, are likely more cost-effective and accessible, thereby providing more opportunities for practice. Moreover, SBT can be effective with low physical fidelity equipment if high psychological fidelity is achieved. This suggests that the lower right quadrant of the figure represents a sweet spot that training system developers should aim for. However, to capitalize on the effectiveness of creating high psychological fidelity while avoiding the expense of high physical fidelity, trainers must take steps to overcome assumptions about the value of physical fidelity in SBT.

Fidelity and Training Effectiveness

High Physical Fidelity in SBT Is Overvalued and Expensive

There are long-held and widespread beliefs that greater realism in SBT with respect to physical fidelity produces better training outcomes (e.g., Beaubien and Baker, 2004; Champney, Carroll, and Surpris, 2014; Hays et al., 1992; Liu et al., 2008; Salas et al., 1998; Salas, Paige, and Rosen, 2013). These views have been attributed to Thorndike's "identical elements" principle (e.g., Thorndike, 1906, as cited in Liu et al., 2008; see also Champney, Carroll, and Surpris, 2014; Kozlowski and DeShon, 2004) such that minimizing differences between training and performance contexts facilitates learning and transfer of training. However, comprehensive reviews of SBT show that assumptions about the need for physical realism are not supported by empirical research (e.g., Salas et al., 1998; Stewart et al., 2008), and physical realism is sometimes counterproductive. For example, for novices, high levels of physical fidelity can be detrimental to learning because they provide too much information (e.g., see Hays and Singer, 1989; Noble, 2002) (although, as we discuss later in the chapter, motion in simulators can be useful for training novices in some circumstances). Hays and Singer (1989) also point out that some activities, such as stopping the action to provide feedback to trainees, depart from realism but are effective instructional practices.

Much of the research on fidelity in SBT has been conducted in the context of aviation and has addressed both technical skills and crew resource management (CRM), but studies of SBT in other domains, such as health care (Beaubien and Baker, 2004), have reached similar conclusions. The research findings are clear (1) that high levels of physical fidelity do not necessarily produce better learning or transfer of training and (2) that transfer of training can occur with low physical fidelity devices. What matters for designing effective SBT is *fostering psychological fidelity* and *application of established learning principles.*

However, despite decades of evidence to the contrary, assumptions about the need for physical fidelity in equipment design have persisted.

Salas et al. (1998) noted that, although there have been advances in research on learning principles relevant to SBT, such as feedback, measurement, guided practice, and scenario design, there has been limited application of this knowledge to the human side of SBT. Salas et al. (1998) attributed this imbalance between human-centered and machine-centered concerns to two factors: (1) The majority of funding goes to equipment development rather than to advancing knowledge of learning processes; and (2) the development of criteria for designing simulators lacks front-end analysis and fails to consider how users learn, perhaps because of the time and resources required for these efforts. As a result, designers and other stakeholders default to assumptions about the importance of physical fidelity. The sunk costs of the substantial investments made in developing, procuring, operating, and maintaining training devices with high levels of physical fidelity might also discourage consumers (in this case, the Army) from rigorously testing whether the equipment fosters better training outcomes than lower cost approaches would. Salas et al. also noted that, when evaluation of SBT occurs, it tends to focus on the *performance of the simulator* and relies on users' reactions to the system or training rather than on assessing the *performance of the trainees* and use of objective measures of performance or transfer of training (see also Hays and Singer, 1989).

Ten years later, the comprehensive review by Stewart et al. (2008) drew similar conclusions about physical fidelity in SBT. Stewart et al. noted the paradox in the aviation training community, i.e., "the technological base for simulation continues to evolve at a rapid pace, while the training programs supporting them have shown very little change over the past 20 years" (Stewart et al., 2008, p. 16). Similarly, Cannon-Bowers and Bowers (2009) concludes that SBT is typically developed with an emphasis on technology, with little consideration of pedagogy. Moreover, a recent study of virtual training in the Army shows that these issues persist (Government Accountability Office [GAO], 2016), stating that Army regulations and training strategies identify technical requirements for development of virtual devices but provide little specification of tasks to be trained, guidance for how to use the devices to conduct training, or requirements for postfielding analyses of training effectiveness.

An emphasis on simulator characteristics and subjective assessment is also apparent in efforts to determine appropriate levels of fidelity in SBT. The Army and other services have supported development of a number of models or decision aids to assist in training design (e.g., see Goldberg and Khattri, 1987) or device configuration (see Hays and Singer, 1989) to determine the degree to which collective tasks can be performed (and therefore, potentially trained) using a particular SBT system (e.g., Burnside, 1990); identify whether networked simulations can supplement other forms of training to inhibit skill decay in teams (Swezey et al., 1998); and recommend training approaches, such as live, virtual, constructive, or hybrid approaches (e.g., IFC International, 2013; Sticha, Campbell, and Knerr, 2002). Typically, these models decompose tasks, equipment, and/or environments into discrete elements. For example, dimensions of tasks might consist of subtasks; task conditions and standards; or features of tasks, such as difficulty of task activities (e.g., identifying problems, solving problems, following procedures, communicating, making decisions). SMEs rate the discrete elements on multiple criteria as low, medium, or high, and ratings are combined using an algorithm to yield a recommended training approach.[2]

These models are intended for task analysis, which is a critical first step in training design. However, ratings of the model criteria are based on SME judgments and, therefore, are influenced by subjective views and, sometimes, institutional norms about fidelity rather than

[2] For example, in a model from IFC International (2013), collective tasks within different occupational families—such as "transport personnel and cargo" (transportation), "support a water crossing operation" (combat engineering), and "evaluate casualties" (chemical, biological, radiological, and nuclear defense)—are rated on factors pertaining to requirements for teamwork, synchronous activity (i.e., coordination), and environmental conditions (i.e., use of actual equipment; operating in different environments, such as darkness or noise; processing nonverbal cues; and simultaneous engagement in motor tasks). Ratings might be "low," "medium," and "high" or "not needed," "preferred," and "essential." The ratings are plotted on a radar chart and compared to prescribed thresholds or are reviewed to determine whether particular cutoffs apply, e.g., if the need for teamwork training is "high," the recommendation is to train in a live environment, regardless of ratings on the other factors.

by objective measures of learning and training transfer. This issue is less problematic when the criteria are grounded in established empirical findings, but the models vary in the strength of the underlying support for their assumptions, and new research findings and changes in training technologies could require continual updating of model criteria. Other authors also note pervasive subjectivity in decisions about fidelity in SBT and problems with subjective judgments (e.g., Liu et al., 2008; de Winter, Dodou, and Mulder, 2012; Salas et al., 1998). In addition, although considerable effort has gone into development of these models and collecting SME judgments (which can entail making thousands of ratings), we have not found systematic efforts to validate these models (e.g., by comparing how approaches recommended by the models and alternative approaches affect learning and transfer of training) and have not found evidence that such models are being used in practice.

Beliefs that simulators should replicate the physical features and functions of actual equipment impose substantial financial costs and limit training efficiency. Clearly, achieving high levels of simulator realism is costly in terms of system design and procurement. Operation and upgrades also impose costs. Such systems as CCTT and AVCATT are generally housed in large facilities and require contracted personnel with specialized system expertise to run the equipment. In addition to the direct costs of these resources, the need for personnel to set up and run training exercises limits access to equipment, which in turn inhibits opportunities for deliberate, repeated practice—a critical factor in developing expertise (e.g., Ericsson, 2006; Ericsson et al., 1993; see also Staller and Zaiser, 2015). Beliefs about the importance of replicating physical features of training systems also create a continuous need to reconfigure and update simulators to match the features of fielded equipment. Thus, when physical realism drives the acquisition process, rapid changes in technology quickly render simulators obsolete, a cycle that has been likened to the "quest for the Holy Grail" (Stewart et al., 2008, p. 6). In a series of focus groups, interviews, and surveys addressing simulations for collective training (Seibert et al., 2012), SMEs indicated that the lack of concurrency between simulators and real aircraft raises the possibility of negative transfer, i.e., that training in an outdated simulator will impair performance in the actual aircraft. The

SMEs also reported that a lack of concurrency can interfere with collective training because it requires experienced pilots to recall how to operate the outdated systems when training in the simulators.

Psychological Fidelity in SBT Is Critical for Collective Tasks and Is Less Costly but Undervalued

As discussed in Chapter One, collective training rides on the assumption that individual team members and individual teams are proficient in their technical areas of expertise. For example, a collective training exercise for ground maneuvers using CCTT might entail distribution of direct fires among the four tanks in a platoon or among multiple platoons to ensure complete coverage of enemy targets and engaging the most dangerous threats first. The focus of this training is on coordination among the tanks or platoons in terms of such behaviors as monitoring others' actions, providing back-up when others are overloaded, and synchronizing actions (e.g., see Ford and Schmidt, 2000; Rosen et al., 2012; Stewart et al., 2008). The need for these behaviors is intensified in military decisionmaking contexts and in other domains, such as emergency responding, because these situations often lack clear structure; are subject to ambiguous and conflicting information; and require teams to rapidly detect and diagnose problems and implement, monitor, and revise solutions (e.g., Kozlowski and DeShon, 2004). These activities are primarily cognitive (Kozlowski and DeShon, 2004; Stewart et al., 2008), relying on information monitoring and communication to establish situational awareness, develop shared mental models, and coordinate actions—activities that are essential to effectiveness of MTS (e.g., de Vries et al., 2016; Owen et al., 2013). Indeed, literature has shown that, when training such collective tasks, physical and functional fidelity are less important than psychological fidelity (e.g., Bowers et al., 1992).

In addition, cognitive skills decay more rapidly and require more frequent practice or refresher training than psychomotor skills do, and practice can be accomplished with part-task trainers or other equipment without a high level of physical fidelity (Stewart et al., 2008). Practice, accompanied by feedback, is also essential for teams to learn how to adapt to changing situations (e.g., Salas et al., 2009). GFT provides opportunities for repeated practice in different scenarios, allowing teams to apply knowledge and skills to solve novel problems.

Psychological fidelity for collective training can be achieved through systematic analysis of critical task behaviors, creation of scenarios to elicit the behaviors, development of assessment criteria and practices that provide specific feedback to trainees, and opportunities for practice and repetition. The Event-Based Approach to Training (EBAT) outlines a process to achieve these goals, linking training objectives, design, and evaluation (e.g., Fowlkes et al., 1994; Fowlkes et al., 1998; Johnston et al., 1997). EBAT facilitates evaluation, comparisons within teams over time or across teams, and assessment of different performance strategies. In addition, a systematic approach, such as EBAT, can also provide a framework for researchers to identify the most important variables (e.g., task characteristics, training design, sources of feedback, team composition) to manipulate (i.e., the independent variables) to foster psychological fidelity and understand their effects on training outcomes. Use of a framework is important, given that any variables can affect team performance, and it is not feasible to manipulate all relevant constructs simultaneously (Bowers and Jentsch, 2001). A framework can also help researchers identify and control for potential extraneous variables, i.e., variables that can influence the effects of independent variables on outcomes (Bowers and Jentsch, 2001). We describe the EBAT approach in more detail in Chapter Six.

Examples of Studies Examining Low Physical and High Psychological Fidelity SBT for Teams

Hays and Singer (1989) stated that, without research demonstrating the value of low physical fidelity simulations, substantial financial resources would continue to be spent on high physical fidelity simulations. Fortunately, there has been a shift from studying high to low physical fidelity in SBT (Stewart et al., 2008). Here, we provide some examples of studies addressing SBT for cognitive team tasks.

A variety of studies from the literature examining CRM training for flight crews provides support for the ideas that (1) high physical fidelity SBT is not required for team training, and (2) low physical fidelity can be effective when combined with high psychological fidelity. CRM uses detailed scenarios to simulate high levels of psychological fidelity for crew planning and execution tasks, including criterion stressors such as time pressure and information overload. Physical fidelity in CRM training might be extremely low, e.g., consisting only of a desktop computer and multiple monitors (Prince and Jentsch, 2001). For example, Stout et al. (1998) described a program of research investigating low physical fidelity simulation for military aviation CRM in two-person crews. The equipment consisted of two networked desktop computers and communications via intercom. This training used interdependent tasks, emphasizing team skills, such as mission analysis, communication, leadership, adaptive performance, situational awareness, and shared decisionmaking. By and large, these studies showed that low fidelity training promotes CRM processes relative to a no-training control group and provide evidence for the association of CRM processes with task performance. Similarly, in a review of 58 studies of CRM training, Salas et al. (2001) concluded that findings support the use of low-physical fidelity simulators for CRM training and that CRM training has led to positive attitudes, learning, and behavioral changes on the job.

Toups et al. (2011) conducted a study of zero-fidelity simulation for emergency response teams (ERTs), which, like teams in military contexts, operate under high levels of stress and make decisions that have life-or-death consequences. Zero-fidelity simulations consist of abstractions of task requirements; i.e., the simulations evoke the cognitive, affective, and behavioral requirements of a task but do not simulate the concrete or real-world aspects of the task or environment. In Toups et al. (2011), equipment consisted of laptop computers, game software, communications tools, and physical space to arrange players in a distributed fashion. The task consisted of a context-free game that was analogous to an ERT in terms of team member roles; time pressure; and needs for information processing, communication, and coordination. Thus, although the simulation had very low physical fidelity, it offered high psychological fidelity. Quantitative analysis of team communication and game logs showed that teams demonstrated improved coordination and communication in training, and qualitative data provided evidence of engagement and transfer of training to ERT performance.

Recently, there has been burgeoning research on SBT in health care settings. As with emergency response and military teams, members of health care teams typically have different areas of expertise, and teams operate in ambiguous, dynamic environments in which problems may have multiple possible solutions and require rapid decisionmaking. Studies in health care have come to many of the same conclusions as research in aviation and other military tasks or settings regarding assumptions about physical fidelity and the need to apply learning principles to the design of SBT (e.g., Benishek et al., 2015; Graafland, Schraagen, and Schijven, 2012; Hamstra et al., 2014; Norman, Dore, and Grierson, 2012; Rosen et al., 2012; Salas, Paige, and Rosen, 2013; McRobert et al., 2013).

Finally, a critical use of SBT is to train individuals and teams to perform under conditions of stress. The kinds of processes required for performance in military teams (information monitoring and communication to develop situational awareness and shared mental models) are also critical for team adaption to stressful conditions (e.g., Entin and Serfaty, 1999). Decades of research show that creating stressful conditions does not require physical fidelity. A recent review of research in law enforcement emphasizes the fidelity or realism of training scenarios over physical realism for training officers to perform under stress (see Staller and Zaiser, 2015).

We do not attempt to review the enormous literature on stress induction and management; instead, we limit this discussion to a small number of illustrative publications, coupled with findings from interviews we conducted in for this research (described in Chapter Four). A number of findings relevant to team performance under stress come from the Tactical Decision Making Under Stress (TADMUS) program, sponsored by the Office of Naval Research. TADMUS examined the nature of stress in tactical crews, the effects of stress on decisionmaking, and strategies to mitigate stress including training and design of information displays (e.g., Cannon-Bowers and Salas, 1998). The TADMUS program distinguished two categories of operational stressors: (1) *Task-related stressors*, which are inherent in the task, such as workload, time pressure, information uncertainty, and auditory overload, and (2) *ambient stressors*, which are in the environment, such as auditory or visual distractions, performance pressure (e.g., from the commander), and fatigue due to sustained operations.

Stressors from both categories can be simulated—whether teams are being trained in simulators with high or low physical fidelity—through the use of *scenario-based training*. That is, task-relevant stressors can be produced in scenarios in numerous ways—manipulating the number of tasks trainees must complete, unexpected changes in task requirements, the time available for task completion, the amount and ambiguity of information about hostile contacts or environments, the number of communication or visual channels that trainees must monitor, the risks associated with failure, perceived threats, and so forth. Ambient stress can be created by manipulating background noise or visual distractors, the length of the training exercise, temperature in the simulator or facility, amount of visual information (too little or too much), and amount of physical space (e.g., Cannon-Bowers and Salas, 1998; Cohen, Brinkman, and Neerinex, 2015; Driskell, Johnston, and Salas, 2001; Schnell, Postnikov, and Hamel, 2011; Schnell et al., 2012). As we discuss in Chapter Four, the use of hybrid approaches that mix VME with inexpensive physical environments (e.g., plywood enclosures) can simulate confined space within the vehicle to induce stress.

We acknowledge, however, that some stressors cannot be replicated in SBT or require extremely high levels of physical fidelity that are not present in AVCATT, CCTT, and current VME approaches. For example, the threat of death or injury can be experienced only in live fire training or in actual combat. Replication of multiple degrees of vehicle movement can influence even simple acts such as flipping one's helmet switch to talk on the radio. In live training, or in a simulator with very high levels of physical fidelity, vehicle commanders would have the opportunity to learn how and when they need to halt their vehicle or modify actions to merely speak on the radio. We also note, however, that this level of fidelity is very expensive to produce, and past studies examining multiple degrees of motion in aviation simulation training have shown mixed results.[3] In addition, these findings pertain to individual training for technical skills—skills that operators are expected to have mastered prior to collective training events.

Taken together, the vast research on physical and psychological fidelity in SBT suggests that, if training design is sound, training for many collective tasks should be equally effective when using lower or higher physical fidelity equipment, particularly for cognitive tasks. Nonetheless, we have not found studies examining fidelity in SBT for collective training (i.e., MTS) or studies of what might be considered the definitive question regarding fidelity in SBT for military collective tasks: comparing the effects of collective training using PSME (CCTT, AVCATT) and VME on learning and training transfer. Conducting such studies poses a number of challenges, which we describe in Chapter Six.

We also note that findings about the value of physical fidelity are mixed with respect to learners' levels of expertise. Some studies of simulator motion on technical performance have found that motion is important for training individual operators with lower levels of expertise but not for experts (de Winter, Dodou, and Mulder, 2012). However, as noted earlier, others report that high levels of physical fidelity create information overload for novice trainees and are therefore detrimental for performance (Hays and Singer, 1989; Noble, 2002). Given that members of a team or MTS must maintain their individual technical skills and that individuals use simulators to do so (Seibert et al., 2012), more research is needed to understand the association between physical fidelity and levels of expertise for individual training

[3] For example, early studies by Martin and Waag (Martin and Waag, 1978a; Martin and Waag, 1978b) showed that motion did not affect transfer of training. More recently, a meta-analysis of 24 studies (de Winter, Dodou, and Mulder, 2012) found an overall positive effect of simulator motion on transfer of training, but the effect was moderated by several important factors. For example, the effect was small for studies comparing true training transfer (in which the transfer task is conducted in a real aircraft) to quasi-transfer (i.e., in which the transfer task is conducted in a simulator). Effects were smaller for fixed-wing aircraft than for helicopter and disturbance tasks (motion from external forces, such as wind shear or engine failures, as opposed to motion arising directly from the pilot's actions). Motion was of no benefit for training expert pilots but did benefit study participants with no flight experience or intermediate experience levels.

and whether the level of physical fidelity matters for training MTS in which individuals or teams vary in their level of expertise with respect to *collective skills*, which involve perceptual, decisionmaking, communication, and coordination activities.

Summary

Simulators with high levels of physical fidelity are not required for effectively training collective skills, impose substantial costs, and may limit opportunities for practice. In contrast, existing research suggests that lower physical fidelity simulations, such as VME, can be effective for collective training when the training is designed to provide high levels of psychological fidelity; are likely more cost-effective; and can be more accessible, thereby providing more opportunities for practice. The overvaluation of high physical fidelity and undervaluation of high psychological fidelity are barriers that must be addressed in any plan to increase reliance on VME systems as the Army transitions to the STE. Chapter Six includes several recommended strategies for overcoming these barriers.

In Chapter Four, we discuss findings about Army stakeholders' views of the need for physical fidelity in SBT.

CHAPTER FOUR

Stakeholder Views of Collective Simulation-Based Training Systems

Although the research literature surveyed in Chapter Three provides solid scientific evidence regarding the effectiveness of simulation-based collective training technologies, such evidence alone is not sufficient to guide policy and practice. It should be augmented and contextualized by the experience-based perspectives of stakeholders involved in the virtual training of collective skills at the platoon and company levels. Through the design, delivery, and assessment of training, stakeholders develop potentially important insights into the relative values, strengths, weaknesses, and risks associated with training systems, as well as possible changes to the delivery of collective armor and aviation training.

We sought such insights through interactions with individuals representing several key stakeholder groups for Army SBT, including capability providers (e.g., training developers and research staff), training providers, and training consumers (see Table 4.1). We used interviews and focus groups to understand the range of stakeholder perspectives on SBT. We used the stakeholder insights, in combination with findings from the research literature, to develop and field a survey to quantitatively assess stakeholders' beliefs about and preferences for SBT.

In this chapter, we first describe the method used for interviews and focus groups, followed by the themes extracted from their answers to interview questions. We then describe a survey of consumers of collective SBT systems implemented at the U.S. Army Aviation Center of Excellence (USAACE) for aviation training.

Table 4.1
Interview and Focus Group Sites and Stakeholders

Site	Stakeholders	Number of Participants
TCM V&G	Capability providers: training developers, research staff	4
USAACE, Fort Rucker	Training providers and consumers: schoolhouse leadership, SMEs, warrant and commissioned officers	10
U.S. Army Maneuver Center of Excellence (MCoE), Fort Benning	Training providers and consumers: schoolhouse leadership, SMEs, commissioned officers	10
Mission Training Complex	Training providers: PEO STRI staff, SMEs	6
Two ABCTs at Fort Hood	Training consumers: battalion leadership, staff officers, master gunners, company commanders, noncommissioned officers (NCOs), and enlisted soldiers	29
Combat aviation brigade	Training consumers: warrant and commissioned officer aircrews	4
UK Ground Forces Land Warfare Center (Warminster, England)	Capability and training providers: developers, site managers, and staff	5

NOTE: Some numbers are estimates based on discussions with soldiers and contractors during tours of sites and breaks in training.

Interviews and Focus Groups with Stakeholders

Method

We conducted site visits to gain access to SMEs and members of stakeholder communities. We visited a range of sites and conducted interviews and focus groups with stakeholders representing diverse perspectives, as shown in Table 4.1. In total, we met with approximately 68 respondents. We conducted focus groups with brigade- and battalion-level staffs involved in development, execution, and assessment of collective SBT. The remaining discussions consisted of one- or two-person interviews.

We used a semistructured format for interviews and focus groups. Appendix B provides the interview and focus group questions. During site visits, not all topics were covered with all respondents because some discussions were carried out during breaks in training or during a tour of a facility. The discussions covered issues related to CCTT/AVCATT and GFT fidelity, value for collective training, facilitators and barriers to use, and recommendations for improvement. Most interviews occurred in scheduled meetings, and others occurred during facility tours or with soldiers during breaks in their training. We then discussed and synthesized participants' responses to extract main themes.

Results

Two primary themes emerged from the interviews and stakeholder's comments on these topics. One concerns fidelity and currency of simulators, which is closely tied to perceptions of the value of simulators for collective training. The second concerns factors that facilitate or inhibit use of the systems.

Fidelity, Currency, and Perceived Value of Simulators

PSME

Many stakeholders from each group reported valuing high physical fidelity simulators. Stakeholders generally reported that CCTT and AVCATT are valuable and important for platoon- and company-level collective training, although there was some variation in views about the degree to which current PSME meets expectations for fidelity and in perceived value. Responses about PSME ranged from strong beliefs in the value of the systems (e.g., "Don't take CCTT away") to weaker beliefs ("Use it if there's nothing else to do"). Training consumers reported that CCTT is particularly useful for collective tasks in which maintaining and communicating situational awareness are fundamental to task accomplishment, such as platoon maneuver and distribution of direct and indirect fires during engagements. Stakeholders also reported that CCTT and AVCATT are helpful for learning individual and crew skills, but these skills are beyond the scope of this research. In addition, individual and crew training can be accomplished using other TADSS, ranging from static cockpits to full-motion simulators, and the Advanced Gunnery Training System trains gunner-tank commander teams.

A number of respondents from the stakeholder group discussed issues with physical simulators in terms of lack of concurrency. They reported that PSME lags the changes in actual military equipment platforms. For example, there were reports of still having PSME of outdated Operation Desert Storm model M2 Bradley Fighting Vehicle and outdated models of AH-64 attack helicopters. Unit commanders, SMEs at MCoE and USAACE, and training providers at both reported not wanting soldiers to train on physical simulators that do not match the model and equipment of their standard unit equipment because of concerns about negative transfer of training (i.e., behaviors learned in the simulator will lead to errors when operating actual equipment because of differences in the two platforms). These concerns are consistent with findings of Seibert et al. (2012) regarding the potential for negative transfer resulting from a lack of concurrency in simulators and aircraft. ABCT training consumers and Mission Training Complex staff also reported that crews do not fight in CCTT the same way they do in actual armor at the NTC; they are "buttoned up" (i.e., inside a closed manned module) in CCTT and out of the turret at NTC. As a result, the tank commander cannot get the situational awareness in the same way in CCTT that he or she would when training at NTC. In comparison to GFT approaches, consumers reported that the physical fidelity in CCTT generates more realistic conditions of stress by requiring crews to operate in cramped quarters for long periods.

Both training capability and training providers tie issues of PSME equipment concurrency to the costs to upgrade the training equipment; they noted that upgrades to PSME are costly, and there is often limited or no funding for these updates. Furthermore, maintaining concurrency is complicated from an administrative standpoint because upgrades to simulators are reportedly not funded by same sources as upgrades to weapon systems. Interviews with stakeholders in the British Army face the same challenges with the pace and costs of upgrading physical simulators.

GFT Approaches

Capability developers generally had favorable comments about fidelity of GFT approaches. For example, they reported that advances in immersive visual displays for training are rapidly increasing in quality, decreasing in price, and providing an increasing sense of physical presence. Moreover, upgrading the digital simulations that comprise the VME of GFT approaches is generally much less costly than updating PSME. Training capability developers also commented on the promise of hybrid approaches, which mix low physical fidelity simulators and immersive visual fidelity systems.[1] In fact, the British Army is realizing such gains; dismounts in armored personnel carriers have PC workstations detached from the PSME and can maneuver separately once they virtually dismount. This integration provides greater breadth of combined arms training than the capabilities of CCTT and AVCATT allow. Similarly, British units play many aspects of missions, including simulated tactical operation centers for mission command, simulated and virtual irregular combatants, high-value targets, and civilians on the battlefield.

Training developers and consumers were more varied in their opinions about fidelity and value of GFT approaches; some of these stakeholders are highly enthusiastic, but others took a wait-and-see approach.[2] Although our interviewees did not have experience with VBS3, some company-level personnel with tank gaming experience reported that GFT can train skills for using terrain (e.g., "berm drills" and avoiding "skylining") and maneuvering around obstacles. Personnel at the USAACE reported that collective tactics could be trained on low physical fidelity systems using COTS hardware, especially with new warrants and officers without significant experience in their air-

[1] Hybrid approaches can mix PSME and VME in two main ways. First, VME (e.g., VBS3 with head-mounted displays to immerse the trainees in virtual environments) can be integrated with inexpensive physical environments (e.g., plywood enclosures) to simulate the confined space within the vehicle to add physical realism and, potentially, induce stress. Second, subsets of a vehicle can be simulated as PSME and integrated with VME; for example, GOTS PSME could be used for sights and turret control mechanisms, as the tabletop Conduct of Fire Trainer for the M2 Bradley Fighting Vehicle does, and for the tank commander and gunner, but these could be driven by PCs and directly integrated into VBS3. In this case, the trainees are interacting with physical systems but accessing a virtual world.

[2] At the limited number of sites we visited, we did not encounter soldiers who had experience using VBS3 via Army hardware suites. This likely is because the GFT program is new relative to other SBT systems, which have been in use for decades.

frames. That is, initial experiences with new crews in reconnaissance missions with VME (via laptops using COTS controllers) suggest that inexperienced crews find value in the system, but more-experienced crews are not willing to use it. We also found that training consumers with experience playing commercial simulation games had much more favorable views of GFT approaches. While we found few aviators who reported use of commercial helicopter simulation games, a number of armor enlisted personnel and NCOs who regularly play such games strongly believed in the value of games to teach collective skills and induce some amounts of stress. They agreed that a GFT approach using laptops and COTS could train many cognitive, perceptual, and communication skills that are critical to collective tasks. However, they felt that games cannot provide the physical stress associated with the confined environments in combat platforms.

Facilitators and Inhibitors to System Use

Interviews with and focus groups of training providers and consumers identified a number of factors that facilitate use of PSME or VME approaches. Although most comments pertained to CCTT and AVCATT, many of the facilitating and inhibiting factors are the same regardless of simulation system, so we do not present results separately for PSME and VME.

Stakeholders recognized that utilization of CCTT and AVCATT is driven primarily by four factors:

- **Command emphasis**—what the commander views as important, either in terms of what the commander encourages or requires. This is expressed in many ways, ranging from communication in informal, daily interactions with junior leaders and soldiers to formal policies and orders. For example, high usage rates at a specific installation, as shown in Chapter Five, were attributed to the division commander's emphasis during the reporting period. Another example is use of the DSTS: At one site, it was reported that use was driven by directives from leadership. Respondents from each stakeholder group also reported that usage of CCTT would be higher if it were required for training qualification. The variations in use across sites and organizations suggest that differences in command emphasis can help push use of training resources.

- **Lack of formal requirements or "credit" for training carried out in these systems.** Stakeholders report that that few commanders require use of CCTT or AVCATT. A common theme in our discussions was that the responsibility for training design and implementation belongs to the commander; Army doctrine and culture place responsibility for training readiness on individual leaders.
- **Lack of training time** on unit calendars during normal home-station operations and schedules. Numerous, competing tasks keep units from taking greater advantage of CCTT and AVCATT. Similar comments were made in interviews with personnel from the British Army.
- **Lack of training in the use of SBT among leadership** at the company, battalion, and brigade levels to ensure training effectiveness. "You get out of it what you put into it" was a repeated sentiment from stakeholders, and training providers and consumers reported that leaders need better training management skills to understand and fully leverage the training capabilities of CCTT and AVCATT. Providers reported that company leaders do not adequately prepare for or manage training, such as executing detailed and useful AARs, because they lack the knowledge, skills, and/or time. Moreover, capability providers reported that numerous TSPs have been developed to support training using CCTT, but the TSPs go unused and "sit on a shelf collecting dust." Commanders sometimes have soldiers use CCTT during open time on their calendars by providing loosely structured scenarios accompanied by informal AARs, which one SME described as amounting to "digital day care." Training consumers reported that units should have a brigade- or battalion-level expert to guide company commanders on best practices for SBT; personnel in this role would be comparable to a "master gunner" with specialized training in SBT and an additional skill identifier, or a "digital master gunner."

In addition to these four factors, access and ease of use can facilitate increased use of training systems. Stakeholders reported that training using CCTT and AVCATT is easy to schedule and can be tailored to the needs of the unit leaders. At the same time, training providers stated that modifications to simulators and software are done by contractors, and software is proprietary, which can limit opportunities for commanders to customize training events. Note that this is also the case for VBS3; use of this system requires reserving GFT sets and having contractors set up and run the training. However, the nature of the COTS hardware for delivery of VBS3 or other VME-based training suggest that there is less need for hardware and network support personnel.

Across stakeholder groups, there was the perception that GFT approaches could provide greater access to training opportunities, especially if sets were provided directly to companies, battalions, or brigades.[3] As mentioned earlier, greater access to PC-based SBT, at lower costs, could potentially be provided by a more soldier-enabled approach to accessing GFT sets and VME. The system could be made more turnkey by leveraging the ease of use and reliability of COTS hardware and networks coupled with the development of software tools to support multiuser gaming, including authoring and feedback tools.

Interviews revealed mixed beliefs in the value GFT-like approaches to allow integration among units above the company level (sister companies, battalions and brigades) to engage in simulated training. Some respondents believed that, as with the use of gunnery scores to publicly acknowledge the gunnery expertise of tanks and platoons, having competitions using GFT could motivate practice and skill building. Others did not believe that there would be such value. In the community of designers and providers, there was some belief in the value of using data from collective skill training via both CCTT/AVCATT and GFT-based systems to inform training goals and scenarios at combat training centers (CTCs).

---

[3] Some unit personnel cited the value of controlling the PSME system trainer Advanced Gunnery Training System for M1 tank gunnery training at the brigade level to provide greater access and increase its use by ABCT personnel. These personnel felt they would get similar benefit from having more unit-based access to GFT sets (currently with VBS3).

Several of the themes from interviews and focus groups were echoed in the results of a survey of end users, which we discuss next.

Surveys of Training Consumers

We developed two versions of a survey to assess the attitudes toward and experiences using collective SBT systems of one group of stakeholders: consumers of training. The survey items were based on findings from the interviews and focus groups, coupled with findings from the literature review. One version of the survey focused on CCTT and VBS3; the other version focused on AVCATT and VBS3. The survey asked participants whether they were familiar with each of these training systems and, if so, about their level of familiarity (i.e., whether they have heard about it, observed training, participated in training, and/or planned training using the system). Subsequently, participants who were familiar with the training system were presented with a series of items about system fidelity, value, access to or quantity of training, AAR support, and other factors that facilitate or inhibit use of the system. The survey also included general items about use of other methods for collective training (e.g., "chair drills," procedural trainers), general questions about perceived requirements for fidelity in training simulations, how respondents learned to plan virtual collective training events, gaming experience, demographic characteristics and job experience, and views on how to improve the effectiveness of virtual collective training. The survey took 15 to 20 minutes to complete.

To recruit survey participants, we targeted students enrolled in the Captains' Career Courses at MCoE and USAACE for the CCTT/VBS3 and AVCATT/VBS3 surveys, respectively. Unfortunately, we were not successful in having course personnel at the MCoE recruit participants for the CCTT/VBS3 survey. At the USAACE, 37 students completed the survey; the number of students invited to take the survey could not be obtained, but course personnel estimated that it was 105 students, indicating a 35-percent response rate. Appendix C presents the AVCATT/VBS3 survey.

AVCATT/VBS3 Survey

Ninety-seven percent of respondents were captains. All participants were in the AC and were in the aviation career field. The average age was 28.47 (standard deviation [SD] = 3.24), and the average number of flight hours was 690.17 (SD = 564.70).[4] Fifty-seven percent of respondents had five or fewer years of experience in the Army; 32 percent had six to eight years, and 11 percent had ten or more years. Figure 4.1 shows the breakdown of number of combat tours and CTC rotations.

Table 4.2 shows respondents' familiarity with AVCATT and VBS3. The majority of respondents were familiar with and had planned or participated in AVCATT training; the majority of students were not familiar with VBS3. These results for VBS3 may not be surprising, given that this system is used largely for ground operations; however, the results might be due to the way the question was phrased, which asked specifically about VBS3 (i.e., respondents might have used the

Figure 4.1
U.S. Army Aviation Center of Excellence Survey Respondents' Experience: Combat Tours and CTC Rotations

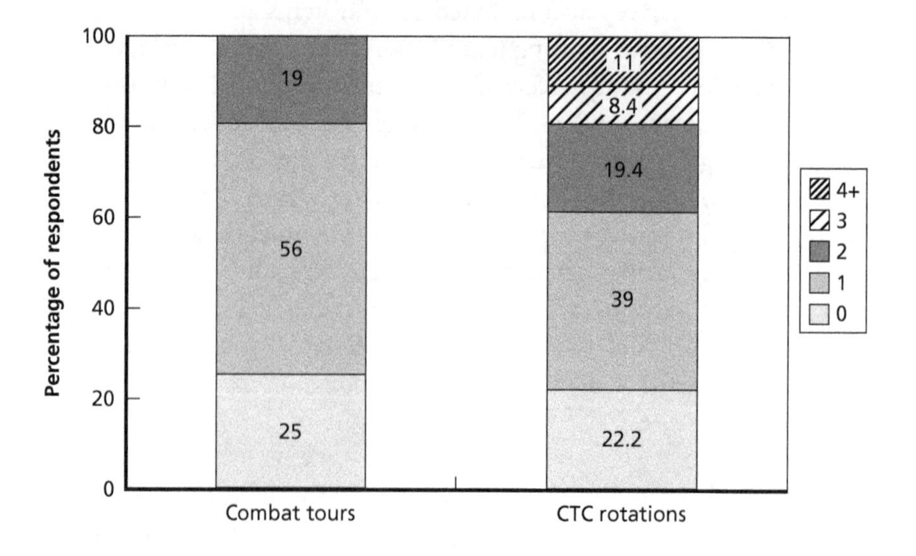

[4] One respondent was an outlier in terms of flight experience, with 3,500 hours. Average flight hours excluding this participant were 607.53 (SD = 286.82).

Table 4.2
Familiarity with Aviation Combined Arms Tactical Trainer and Virtual Battlespace 3 Among U.S. Army Aviation Center of Excellence Students

	AVCATT		VBS3	
Question	**Number**	**Percent**	**Number**	**Percent**
Not familiar with the training system	6	16	29	78
Heard about the system or observed training using the system	6	16	3	8
Participated in and/or planned training using the system	24	65	4	11

NOTE: Number of respondents giving an answer and percentage of all respondents. Percentages do not sum to 100 because of missing values.

system without knowing that it is called VBS3). However, we were surprised that captains in the aviation career field reported not being familiar with AVCATT.

Of the seven students who were familiar with VBS3, only four students answered questions about VBS3 system fidelity, feedback, and ease of use. Given the very small sample, we did not analyze responses to items about VBS3.

We analyzed respondents' answers to attitudes about AVCATT for collective training. Each item consisted of a single statement (see Table 4.3). Response options ranged from 1 = strongly disagree to 4 = strongly agree, along with "unable to judge"; unable-to-judge responses were excluded from analyses using average ratings. Responses were scored so that higher ratings reflect more positive attitudes. For ease of interpretation, we categorized most of the items as pertaining to system fidelity, training quantity, performance feedback, and ease of use. We calculated the average rating across the items within each of these scales or categories. We analyzed some individual items as well (see Table 4.3).[5]

---

[5] Categorization of these items was based on subjective evaluation. A much larger sample of survey respondents is needed to use statistical analyses to determine how items should be grouped into categories.

Table 4.3
Survey Items About Attitudes Toward and Experience Using the Aviation Combined Arms Tactical Trainer

Category	Number of Items	Example Items
Fidelity	16	• Replicates effects of environmental conditions (e.g., rain, snow, wind, dust, and night or day) on aircraft, sensor, and weapon system performance • Creates realistic types of combat stress for collective task training, for example, sensory or cognitive overload or disorientation • Provides high-fidelity visual scenes to support collective task training • Provides accurate weapon and sensor system performance for collective task training • Provides accurate flight model performance for collective task training
Value	2	• Is valuable for preparing personnel for collective tasks during CTC rotations
Access to and quantity of training	4	• Allows units to conduct many iterations of collective task training in a short time • Is available for collective task training when needed
AAR support	3	• Enables useful AARs for collective task training
Ease of use—positively worded	2	• Facilitates rapid planning and preparation of collective task training events
Use for novice personnel	1	• Is effective for collective task training for novice personnel
Use for experienced personnel	1	• Is effective for collective task training for experienced personnel
Command emphasis	1	• My supervisor thinks use of AVCATT is important for collective training

Figure 4.2 shows the average scores for the categories or items shown in Table 4.3. Because most of these measures consist of multiple-item scales, we report average scores rather than frequencies. Average responses for most scales or items ranged from 2.5 to 3.5.

Figure 4.2
Attitudes Toward and Experience with Aviation Combined Arms Tactical Trainer

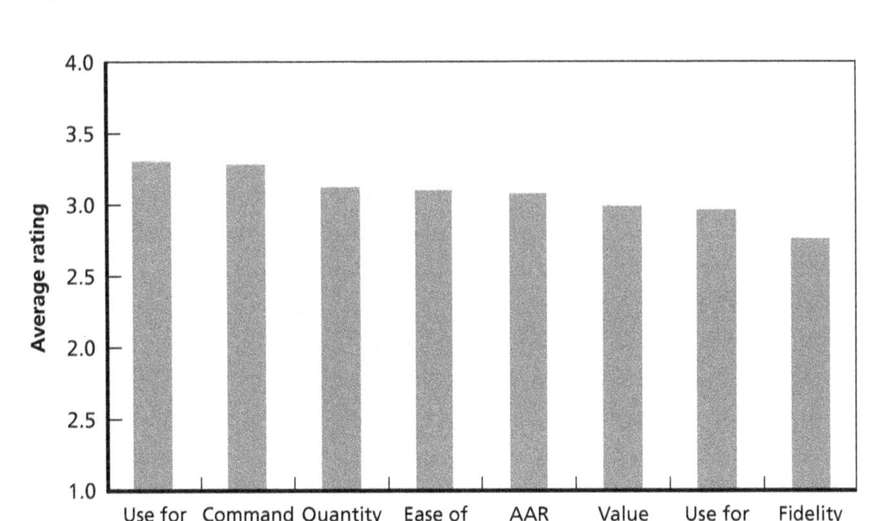

Figure 4.3 shows frequencies of responses for the following single items about simulator use in general:

- Desktop simulators can enhance the effectiveness of live training for mission-critical collective tasks.
- Physical simulators can enhance the effectiveness of live training for mission-critical collective tasks.
- Simulators require high visual fidelity to effectively train mission critical collective tasks.
- Simulators require high physical fidelity to effectively train mission-critical collective tasks.

Responses to the items in Figure 4.3 are presented from most to least favorable. The majority of respondents agreed or strongly agreed with all four statements, although attitudes toward physical simulators were more favorable than were attitudes toward desktop simulators for enhancing live training. Thirty percent of respondents disagreed or disagreed strongly with the assertion that desktop simulators can enhance live training; only 3 percent disagreed (strongly) that physical simulators enhance live training.

Despite the perceived importance of physical fidelity in simulations, as shown in Figure 4.3, ratings of the fidelity of AVCATT were somewhat low, with an average score of 2.74 out of 4.00 (see Figure 4.2). Moreover, items about form and fit (physical fidelity), and functionality (e.g., replicating speed and mobility) had the lowest ratings (2.36 and 2.27, respectively). Ratings for form and fit could be explained by fact that AVCATT is designed to provide training for a number of different rotary-wing aircraft using modular units to physically simulate aspects of the cockpits, and these modular units do not always represent the physical layouts in the cockpit. However, this explanation would not apply to low ratings for the functionality item. The ratings for the remaining items addressing functional fidelity, such as replicating environmental conditions, terrain, visual scenes, and communications capabilities, were higher, ranging from 2.67 to 2.95.

Figure 4.3
General Attitudes About Simulators for Training

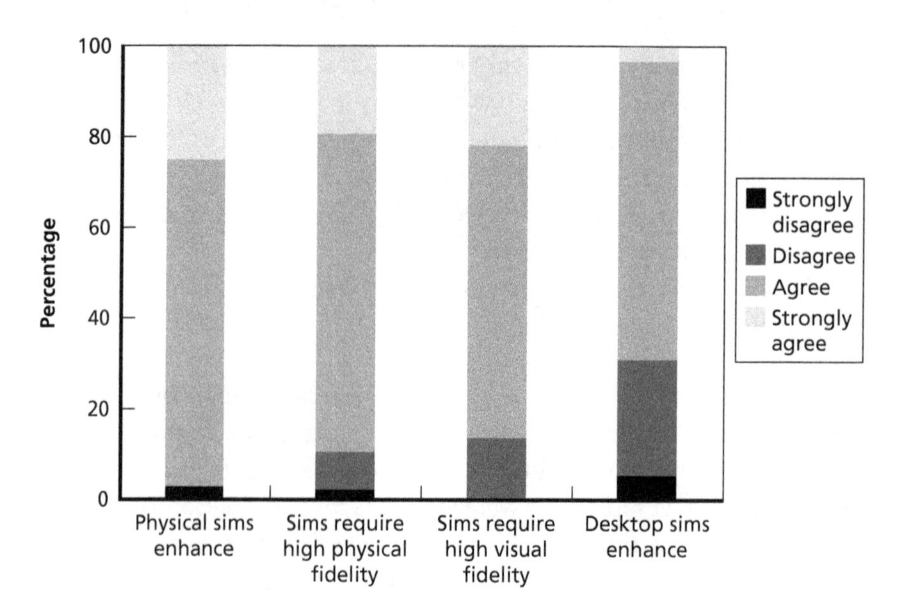

Responses to an open-ended question regarding how to improve the effectiveness of SBT equipment shed light on these ratings about simulator fidelity. Among 20 respondents who answered the open-ended question, seven emphasized the need for congruence between simulators and airframes in use in terms of physical and functional fidelity. Examples of comments include the following:

> Maintain up-to-date simulators. The CH47 in the AVCATT is a D model and is completely different than a [sic] F mode. It changes the reality and effectiveness of the training.

> Virtual collective training needs to be in a simulator whose systems replicate exactly how the aircraft operates and responds in real life.

These responses are consistent with findings from our interviews and focus groups, as well as results reported by Seibert et al. (2012), about a lack of concurrency of simulators and illustrate an issue pointed out in Chapter Three: perceptions about the need to continually update simulators to match current models of military equipment.

We also examined the degree to which views toward AVCATT and general views about simulator training were associated with individual characteristics of respondents. Responses were not associated with flight hours, age, or experience using simulators for training.[6]

Based on findings from the interviews and focus groups, we analyzed the association of responses with gaming experience, with the expectation that respondents with more gaming experience would be less concerned about physical fidelity and have more favorable views of desktop simulations to support training than would respondents with less gaming experience. We measured gaming experience with a 16-item biodata scale asking about participants' experience with video games and related questions about technology use (e.g., I play video games that replicate ground maneuver combat; I discuss games with friends or with others in online forums; I typically modify commercial games rather than playing them as is; I typically have the latest gaming platforms). Response options were "yes" and "no," with scores based

---

[6] In analyses of flight hours, we eliminated responses from the individual who had an exceedingly high number of hours and whose responses had a large impact on results.

on the number of "yes" responses selected. The internal consistency of the scale was high (coefficient alpha = 0.87), and scores on the scale were strongly associated with a separate item in which respondents were asked to report whether they consider themselves to be serious gamers, casual gamers, or not gamers (one person reported being a serious gamer; 15 classified themselves as casual gamers; and 21 classified themselves as nongamers).[7]

Views toward AVCATT and general views about simulator training were not significantly associated with scores on the gaming scale. Comparisons between casual gamers and nongamers were marginally significant for one item, which was the perceived need for physical fidelity in simulations.[8] Surprisingly, casual gamers agreed more strongly with this item than nongamers, with a mean response of 3.29 on a four-point scale (SD = 0.47) for casual gamers and a mean response of 2.91 (SD = 0.70) for nongamers.

In light of interview and focus group findings regarding the need for enhanced knowledge and skills about SBT, we included survey items addressing how soldiers learn to conduct SBT events. Seventeen respondents reported learning through professional military education, such as the Captain's Career Course; ten respondents reported learning through informal, on-the-job training; and three reported learning through a formal class at the Training Support Center or Mission Training Complex. (Some respondents reported more than one of these methods.)

The survey included items about using other methods for training crew and collective tasks, including chair drills for communication skills; training aids, such as engine cutaways or hydraulic boards; and procedural trainers, such as gunnery trainers or cockpit procedural trainers. Figure 4.4 shows the frequency of responses and demonstrates that other TADSS and methods provide value to platoon-level collective skills training for aviators.

---

[7] For the difference between casual games and nongamers in biodata scores, $t_{(14.81)} = 4.67$, $p < 0.001$. Because there was only one serious gamer, he or she was excluded from this analysis.

[8] For the difference between casual games and nongamers, $t_{(33)} = 1.78$, $p < 0.10$. One-hundred percent of casual gamers agreed or strongly agreed with this question, compared to 81 percent of nongamers.

Figure 4.4
Use or Knowledge of Use of Other Methods of Crew and Collective Training

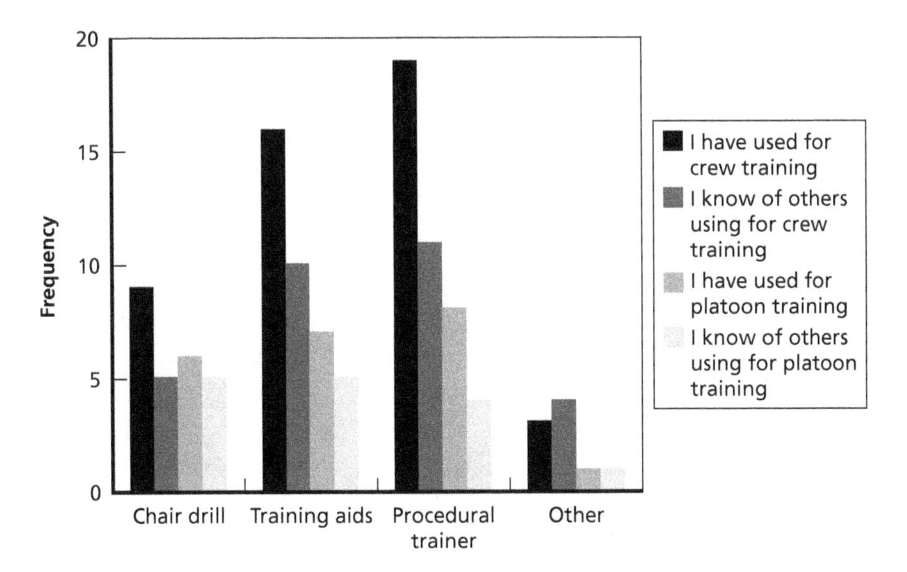

In summary, the survey results, although limited by the sample size, mirror what soldiers in ABCTs and at the MCoE and USAACE told us during site visits: that they perceive high physical fidelity virtual simulators as valuable for training collective skills. However, there was also evidence in both the survey and interview data that desktop or GFT-based simulations can enhance training.

Interestingly, although we found anecdotal evidence in interviews of differing beliefs about the effectiveness of desktop or GFT delivery of armor collective training based on stakeholders' experience with gaming (three enlisted personnel, two NCOs, and one officer out of a total of 29 ABCT soldiers), this was not the case among the aviators who participated in the survey. This could be due to differences in the populations queried; for example, survey participants were officers and may have been older, on average, than were participants in interviews. Or it could be because aviators have similar attitudes toward simulations for aviation training, regardless of their gaming experience. There also was limited variability in the extent to which survey participants were engaged in gaming activities, with most reporting that they are casual gamers. We recommend tracking the perceived value of collective training provided by PSME- and VME-based approaches for armor and aviation consumers of training over time to identify trends in acceptance. The web-based surveys designed for this research could prove to be valuable starting points.

Summary

In summary, interviews, focus groups, and surveys revealed that views of PSME and GFT approaches varied among the diverse stakeholders. Representatives of each group emphasized the value of CCTT and AVCATT for collective training but, at the same time, criticized the lack of concurrency with standard operating equipment and raised concerns about the costs for updating these systems. Attitudes toward the value of GFT approaches (as well as hybrids that combine low and high physical fidelity equipment) were favorable among capability providers and among training consumers who are serious gamers (as assessed in interviews and focus groups) but were more variable among training providers and other training consumers. Key drivers of utilization of all collective SBT systems include command emphasis, policy or requirements for use, commander knowledge and skill in using the systems and related training resources, and accessibility or ease of use. In Chapter Six, we discuss how to capitalize on these drivers to shape future use of collective SBT systems.

CHAPTER FIVE

Comparing the Costs and Cost-Effectiveness of Collective Simulation-Based Training Systems

Cost analysis of training systems, as in other areas, can help the government decide among "alternative strategies, methodologies or settings for achieving certain goals or objectives" (Doughty, 1976). The Army and other government organizations have analyzed the costs of virtual training systems, such as CCTT, in the past, but such organizations as GAO believe that more-comprehensive cost calculations could improve decisionmaking about training in the future (GAO, 2013; GAO, 2016). In this chapter, we explain, estimate, and compare the costs associated with the Army's existing systems for collective virtual training. For consistency, all costs presented in this chapter have been adjusted to FY 2017 dollars. We also estimate and compare the cost-effectiveness of the training systems, although this analysis relies on utilization data and is limited by a lack of data on quality of training. It is worth noting that each of these programs is made up of different modules and was developed to meet different objectives. Therefore, cost comparisons cannot be equivalent until the Army develops common outcome metrics that are independent of the training system used.

It should also be stressed that findings on usage of systems reported here have some limitations:

- AVCATT usage is based on all AVCATT equipment sets.
- CCTT usage includes manned modules in both fixed training sites and MCCTT mobile equipment sets and excludes usage of the RVTT and DSTS.

- GFT utilization data are newer and less reliable than the contractually mandated data collected for AVCATT and CCTT.
- GFT utilization data are not necessarily limited to collective training exercises.

To preview the main findings, our analyses show that the Army's current annual budgeted cost for AVCATT is more than for GFT ($52.3 million versus $36.29 million; see the figures for total cost in Table 5.1). CCTT expenditures are almost twice those for GFT ($65.4 million versus $36.3 million), even though the CCTT program is no longer fielding additional systems. Also, GFT is much less expensive than CCTT and AVCATT once utilization is taken into account, although its utilization rate is more uncertain because limited locations are reporting.[1]

While AVCATT, fixed CCTT, and MCCTT have procurement dollars, they are not fielding additional capacity or locations. These programs are using their financial resources to deal with obsolescence and improve the currency of the existing systems. In contrast, GFT is still in the process of fielding new systems in new locations.

Cost Comparison

Army programs typically have a combination of variable costs and fixed costs. Variable costs are those that vary with operational tempo or utilization, while fixed costs are those that remain the same regardless of output. For the purposes of this analysis, we consider costs that remain steady for one year or more to be fixed. Most costs that we can associate directly with virtual training are fixed. For example, the Army does not pay soldiers more when they train more, so personnel costs are fixed. Existing contracts for contractor support include multiple option years at a steady rate of support, regardless of utilization, and therefore remain fixed for about five years at a time. Current contracts provide a certain level of service month to month, so costs do not vary

---

[1] A 2014 analysis of CCTT and GFT also found that GFT can provide training for an ABCT at lower cost than CCTT (CAC-T, 2014).

Table 5.1
Summary of Findings: Games for Training Is the Least Expensive System, Particularly After Accounting for Utilization

	CCTT[a]	AVCATT[b]	GFT[c]
Total cost ($M)	65.40	52.30	36.30
RDT&E ($M)	0.630	5.08	1.00
Procurement ($M)	43.97	34.65	6.35
Maintenance ($M)	3.47	0.02	0.20
CLS ($M)	17.30	12.51	—
Personnel ($M)	—	—	28.74
Cost per soldier potential training day at current utilization ($ per soldier day)	750.00	7,000.00	200.00

NOTE: Adjusted to FY 2017 dollars.

[a] Average budgeted for FYs 2016–2021. Average actual expenditures for FYs 2010–2015 were lower, at $64.6 million, because of maintenance spending.

[b] Average budgeted for FYs 2016–2021. Average actuals for FYs 2010–2015 were lower, at $39.1 million.

[c] RAND Arroyo Center estimates for supporting personnel combined with GFT budget data.

by level of usage—i.e., the Army does not adjust contracts for virtual training to provide lower levels of service when a unit deploys. Costs of virtual training programs that do vary, such as the cost of electricity usage, are covered by base operating cost accounts that do not allocate specific costs to training facilities (e.g., buildings); as a result, these are excluded from our analysis. In short, from the Army's perspective, the cost of operating a virtual training facility is fixed for several years regardless of the level of utilization.

As Table 5.1 suggests, our cost estimates include five categories of cost:

1. **Research, development, test, and evaluation (RDT&E).** These are appropriations to develop and test new technologies.
2. **Procurement.** These appropriations allow the government to buy systems, upgrade software, modify the displays for concurrency, and purchase new parts for obsolescence.

3. **Maintenance.** These appropriations pay for repair parts and labor for civilian workers involved in maintaining the systems.
4. **Contractor logistics support (CLS).** This is largely an alternative to the maintenance cost category and applies to programs in which a contractor provides personnel and parts to run and maintain the virtual training system. This is the case for CCTT and AVCATT.
5. **Personnel and scenario building**. While GFT does not have centrally managed CLS, locations have contractors and government employees that provide support that are estimated within this cost element.

In the following subsections, we estimate the total costs to the Army of each of the three virtual collective training systems.

CCTT Costs

To estimate CCTT costs, we used program objective memorandum (POM) data, including the 2016 estimate and the 2017 to 2022 submission for CCTT RDT&E, maintenance, and procurement appropriations. Currently, the majority of spending in RDT&E and procurement is on planning for concurrency, dealing with obsolescence, and software upgrades.

In addition, we were able to obtain several years of actual expenditures from the program. While it appears that CCTT is in the midst of procurement, it is only to sustain existing capacity and deal with obsolescence, not to increase capacity. In fact, in 2016, the CCTT program responded to tightening budgets by reducing the availability of some of its training sets, which lowers maintenance and support staff requirements for the contractors at some locations. Therefore, the actual expenditures on CLS and prior utilization for the program need to be put in context. It is possible that future utilization of the facilities will remain the same if sites selected for reduced services were chosen well or if utilization could be reduced because of lower availability, but the impact at this time is unknown. This also means that future total contract costs will be lower, and budgeted CLS is more representative of future costs than actuals. Table 5.2 shows the actual expenditures,

Table 5.2
Average Annual Cost for Close Combat Tactical Trainer

	Actuals FYs 2010–2015 ($M)	Budgeted FYs 2016–2022 ($M)
RDT&E	3.61	0.63
Procurement	38.66	43.97
Maintenance	0.25	3.47
CLS	22.09	17.27
Total	64.62	65.35[a]

NOTE: Adjusted to FY 2017 dollars.

[a] This is the total cost number shown in Table 5.1.

excluding dismounted soldier, for FYs 2010 to 2015 and the POM funding level for FYs 2016 to 2022.

AVCATT Costs

To estimate AVCATT costs, we used POM data, including the 2016 estimate and the FY 2017–2021 and FY 2018–2022 requests for AVCATT RDT&E, maintenance, and procurement appropriations. As with CCTT, AVCATT procurements will not add more sites; rather, these expenditures are just intended to keep the existing sites operational. In addition, we were able to obtain FY 2010–2015 actual expenditures from the program. Given the increase in procurement in the budget years, we anticipate additional investments in software and fidelity but have limited insight into the nature of these investments. The costs from the POM submissions and prior actual expenditures are shown in Table 5.3.

GFT Costs

For GFT, we obtained budget data for the central program from POM submission data for FYs 2016 to 2022, but the program was reluctant to release actual expenditure data. Because the central program does not include all costs for support in the field, we had to make assumptions about the numbers of personnel involved in that effort and their costs based on limited reporting.

Table 5.3
Average Annual Aviation Cost for Combined Arms Tactical Trainer

	Actual FYs 2010–2015 ($M)	Budgeted FYs 2016–2022 ($M)
RDT&E	4.44	5.08
Procurement	19.51	34.65
Maintenance	0.44	0.17
CLS	15.13	12.51
Total	39.12	52.26[a]

NOTE: Adjusted to FY 2017 dollars.

[a] This is the total cost number shown in Table 5.1.

Total costs for GFT are lower than for the CCTT program, which is to be expected, given its smaller equipment footprint and lower physical fidelity. The program was unwilling to share actual expenditures on RDT&E, procurement, and maintenance. The program was completing its planned procurement; thus, the data were sensitive. We would expect the procurement line to decrease in future POMs and the maintenance costs to remain similar or slightly increase. Costs for personnel used to support training exercises and scenario development are not centrally collected or managed because the systems themselves are not centrally managed. The individual sites are not connected, so the system cannot automatically aggregate usage data and cannot easily share scenarios across sites. Budgeted expenditures for the program are available in Table 5.4, combined with our estimates of costs for support personnel and scenario building, calculated at a rate of 1.25 contractors per set.[2] The average number of contractors per set was taken from a limited data set of locations that are directly supported by GFT and not for locations that are unit managed. There is a pos-

[2] The number of personnel per set ranged from 0.50 (at one site) to 2 (at one site), with an average of 1.26 and a standard error of 0.15. Given that there is an average of 137 GFT sets from FY 2016 to FY 2022, the cost of personnel and scenario building in Table 5.4 may be ±$7 million, depending on staffing decisions.

Table 5.4
Average Annual Cost for Games for Training

	Actual FYs 2010–2015 ($M)	Budgeted FYs 2016–2022 ($M)[a]
RDT&E	Unavailable	1.00
Procurement	Unavailable	6.35
Maintenance	Unavailable	0.20
Personnel and scenario-building	Unavailable	28.74[b]
Total	Unavailable	36.29[c]

NOTES: Adjusted to FY 2017 dollars. Average number of sets in the POM from FYs 2016–2022 is 137 at more than 100 locations.

[a] RAND estimates.

[b] RAND estimated the cost of personnel and scenario building because it is not centrally managed by the program.

[c] This is the total cost number shown in Table 5.1.

sibility that unit-managed locations operate differently, but interviews with GFT program officials indicated that they still rely on contractor and civilian assistance. Therefore, our analysis assumes that this support is available, but ours is likely a conservative estimate of costs. Notably, personnel supporting these sets dominate the cost of the program; reductions in such personnel could radically decrease the cost per training, and centralized scenario management could reduce the number of contractors required for each set. The cost of contractors was taken from an earlier analysis and updated with official inflation factors (CACT, 2014).

Cost Comparison

When we roll up the costs across systems, differences in how the programs use funds become clearer (Figure 5.1). CCTT has higher expenditures for fewer sets (seven CCTT and 16 MCCTT) than GFT does. Operations and maintenance costs across the platforms are low in the case of AVCATT and CCTT because of the comprehensive CLS costs and, in GFT's case, because of the simplicity of the hardware systems. CCTT has ongoing expenditures to retain and deal with obsolescence.

Figure 5.1
Close Combat Tactical Trainer and Aviation Combined Arms Tactical Trainer Actuals Compared to Games for Training Estimates

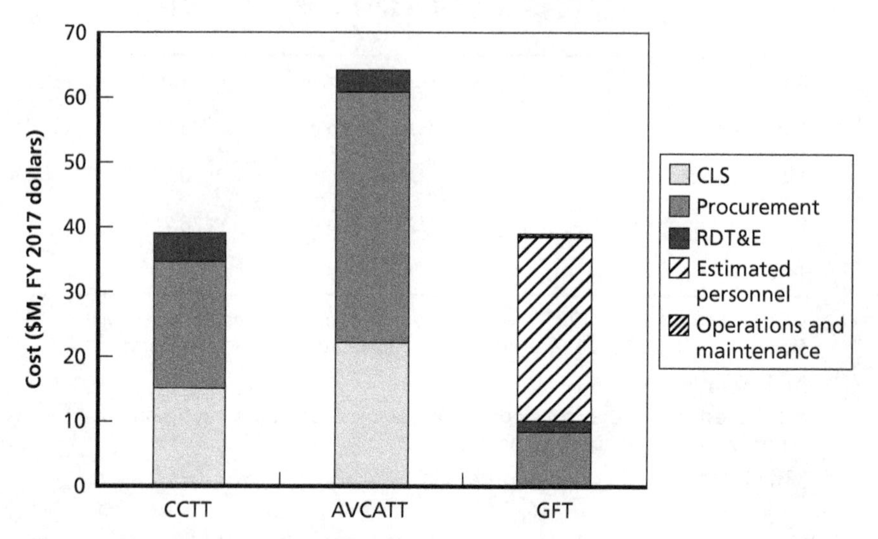

NOTE: As described on p. 62, for CCTT and AVCATT, operations and maintenance and estimated personnel are included in CLS.

AVCATT has similar ongoing expenditures on RDT&E and procurement to stay up to date at its 22 training locations. Comparatively, GFT has low ongoing procurement for software and technical refresh for its 137 sets. Because of the large number of GFT sets, the personnel associated with building scenarios and supporting soldiers in use of the sets is quite high and exceeds the CLS costs for AVCATT and CCTT. As mentioned previously, GFT does not currently have a centralized library of scenarios, and leaders at each location want to develop scenarios specific to their own unit tasks and missions. This reduces the reusability of scenarios across sites.

Utilization Comparison

To assess the cost-effectiveness of training systems, costs should be compared using measures of the quality of the training. However, the

Army does not have a structured method for comparing how well soldiers perform in virtual or simulated training environments and how such training translates to performance in live environments. GAO has recommended that the Army develop such measures of training effectiveness to support assessment of the value of the Army's investments (GAO, 2016).

In the absence of data on quality or learning outcomes, our cost-effectiveness analysis focuses on utilization as a proxy for effectiveness. By combining cost and utilization data, we can estimate how much it costs the Army for a soldier or platoon to attend a training day. We can also estimate how that cost would change if the Army increased utilization of the systems.

CCTT/MCCTT Utilization

We requested utilization data from the CCTT program that are gathered as a contracting requirement. The CCTT metrics are the most complete of the systems we reviewed because they reflect the entire population of training sets. The CCTT/MCCTT program calculates two metrics of interest, one focused on how much the simulators are in use, and one focused on how many soldiers are trained.

In addition to measuring utilization of training modules, the program tracks soldier days in training, a metric of throughput. The soldier-days metric does not count unique soldiers but rather the number of soldier days spent at the facility. A soldier who is present for any part of the morning or afternoon counts as a *half day*, so a *soldier day* does not necessarily mean 8 hours of training time in the facility. Two soldiers who spend 3 hours each at the facility would be counted as 1 soldier day, and 1 soldier spending 2 days at the facility would be counted as 2 soldier days. Given this calculation method, this metric likely overstates the full number of hours of training and does not indicate how many unique soldiers received training.

The utilization rate metric for the CCTT is defined and calculated as follows:

$$Utilization = \frac{Half\ days\ used\ per\ manned\ module}{Available\ half\ days - Maintenance\ downtime}$$

As shown in Figure 5.2, average monthly CCTT utilization varies substantially by location for AC locations with fixed CCTT sites. Utilization rates are calculated by dividing the number of days the facility was reported as being used by the number of days the facility was available for use. We have also included measures of variability in utilization across the months in the periods covered. This is represented in Figure 5.2 by a box-and-whisker graph, with the middle of the box being the median, the top of the box being the third quartile, and the bottom of the box being the first quartile. The whiskers represent the minimum and maximum for each location. Also displayed in Figure 5.2, as blue diamonds, are the average numbers of resident company-level M1 and M2 units at that location during the period. Note that this count is of units with flags at the site; it does not mean that the units were physically present and not on a mission or at a training site.

Understanding the differential capabilities of sites to have high utilization requires knowledge regarding the number of potential

Figure 5.2
Close Combat Tactical Trainer Active Component Sites, Average Monthly Utilization and Average Number of Company-Level Organizations Resident at the Site, FYs 2015–2016

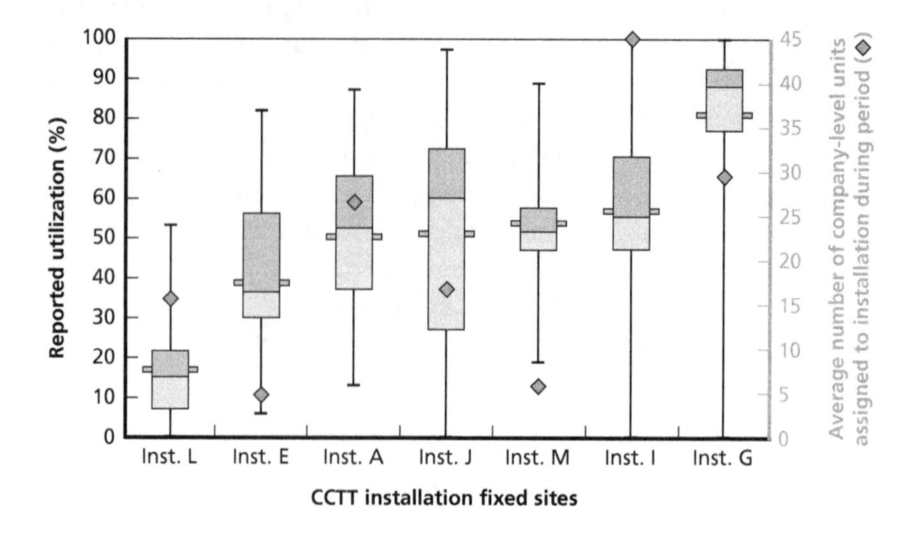

trainee units at a site that are available to train on CCTT. There could be variation across years in the numbers of M1 tank and M2 Bradley companies that were physically resident at home station as opposed to deployed on an actual mission or training exercise. Because this analysis was beyond our original scope, Figure 5.2 includes the average number of M1 tank and M2 Bradley company-level units to serve as a rough surrogate measure for the potential demand for CCTT-based training at each site in FYs 2015 and 2016. Across this two-year period, the variation across installations in usage and number of resident units suggests that the number of potential units does not correlate well with reported usage.

Note that there is a great deal of variation across locations in Figure 5.2, and this variation will also be seen in upcoming utilization reports for MCCTT and AVCATT at AC and reserve component (RC) locations. There are utilization surges, as represented by the maximum height of the whiskers for each location. There are also minimum usage levels in a month, sometimes with zero usage at several sites. This variability in usage could be driven by many factors; for example, higher usage could be a result of

- the total number of units present at the location during the period (compared to units deployed on missions or to a CTC for training)
- preparation time for a mission or training rotation at a CTC (compared to units that had just returned from a mission or training rotation at CTC and are on "Red Cycle," where the training focus is on maximizing self-development opportunities to improve leader and individual task proficiency)
- command emphasis on using collective SBT.

Figure 5.3 illustrates utilization of the mobile MCCTT training sets that ARNG and AC units use.

Fixed CCTT facilities at AC posts have higher utilization than the MCCTTs in trailers, primarily for RC utilization. Given the limited number of training days for RC units, the usage rates are high, but the equipment sets are also smaller, being sized for a platoon rather than the company-plus of manned modules for the fixed sites. These

Figure 5.3
Mobile Close Combat Tactical Trainer Sites, Average Monthly Utilization, FYs 2015–2016

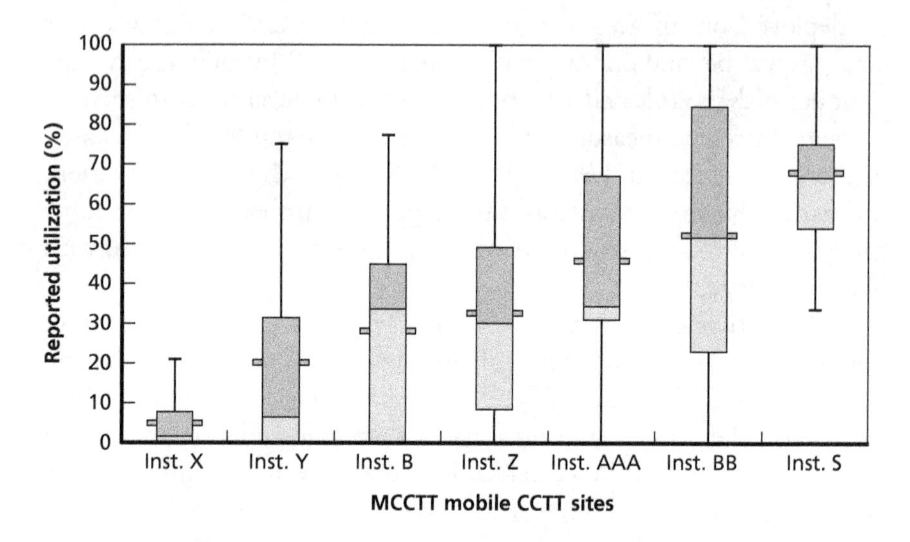

sets are also mobile, so they can be transported to unit locations for scheduled drill days.

As shown in Table 5.5, CCTT utilization across the Army components averaged 38 percent in the most recent contract year.

When we observed a typical training session in the CCTT facility, we learned that only about one-quarter of the time reserved in the facility was actually used for preparation, training, and after-action reporting. The reservation was for three platoons to train for two days for seven hours each day, but only two platoons attended; the third platoon was required for another duty. (This overbooking relative to actual utilization likely occurs across training devices, not just CCTT, including GFT sets.) Across the two platoons, we observed only eight hours per platoon of direct utilization of the PSME-manned CCTT modules. Other time was spent in troop-leading procedures, for example, receiving the mission, developing a plan, and rehearsal; AARs; and other activities, such as meals and transition time. The difference between total reservation time and utilization of the PSME is shown in Table 5.6. To lower costs and make the facilities available to others, the Army should do as

Table 5.5
Close Combat Tactical Trainer Throughput and Utilization, May 2014–April 2016

	May 2014–April 2015		May 2015–April 2016	
	Soldier Days	Average Utilization (%)	Soldier Days	Average Utilization (%)
Active CCTT	67,570	52.60	54,632	47.70
Mobile CCTT	28,589	39.70	31,387	33.10
Combined	96,159	44.50	86,019	38.00

SOURCE: Data supplied by PEO STRI.

Table 5.6
Observed Time Usage of Unit Close Combat Tactical Trainers

	Training	Training and AAR	Preparation, Training, and AAR
Total utilization in platoon hours	8	9.50	11.00
Percentage of reserved hours[a]	19	22.60	26.20
Percentage of hours in facility	72	86.00	100.00

SOURCE: RAND site visit.

[a] Based on 42 hours of time reserved for training.

much preparation without the specialized facility and facilitating labor as possible. This would also be true with GFT: Preparing for training in a separate location before the session with the VME could free time to use the VME. For example, a unit could prepare at the company location before coming to the Mission Training Complex.

AVCATT Utilization

The AVCATT program also aggregates utilization data annually. While there are some data gaps for the last quarter of FY 2016, the data set appears complete. AVCATT appears to have slightly lower utilization than CCTT and GFT, but that may have changed in FY 2016, as shown in Figure 5.4 for AC AVCATT sites and Figure 5.5 for RC sites. This rate of utilization is surprising given the general acceptance

Figure 5.4
Aviation Combined Arms Tactical Trainer Active Component Sites, Quarterly Utilization and Average Number of Company-Level Organizations Resident at the Site, FYs 2015–2016

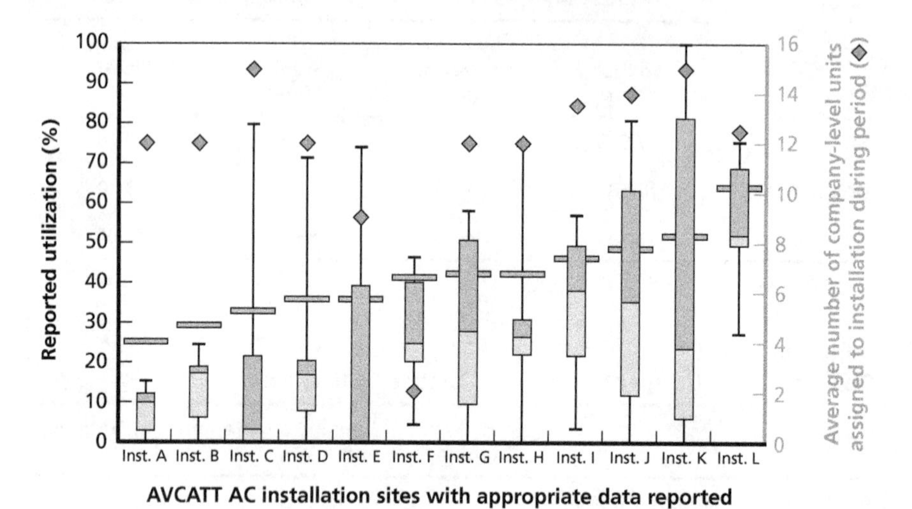

SOURCE: AVCATT FY 2015–FY 2016 utilization as reported to Training Support Analysis and Integration Division (TSAID).

of SBT for other aviation task training generally and requirements for simulation use at some installations.[3] As with the CCTT usage data, we included measures of variability in utilization across the quarters in the periods covered using box-and-whisker graphs. Locations are presented in order of their average usage.

As with CCTT, analyzing the capability of a site to utilize the facility requires knowledge of the number of trainee units available to train on AVCATT. As in Figure 5.2, Figure 5.4 includes the number of aviation companies that were assigned to each AC site, including both AC and RC aviation units and their components. This total can serve as a rough surrogate for the potential demand for AVCATT-based training at the site.

[3] For example, TADSS are reportedly used for 31 percent to 44 percent of required Flight School flying hours at Fort Rucker, varying by airframe type.

Figure 5.5
Aviation Combined Arms Tactical Trainer Reserve Component Sites, Quarterly Utilization, FYs 2015–2016

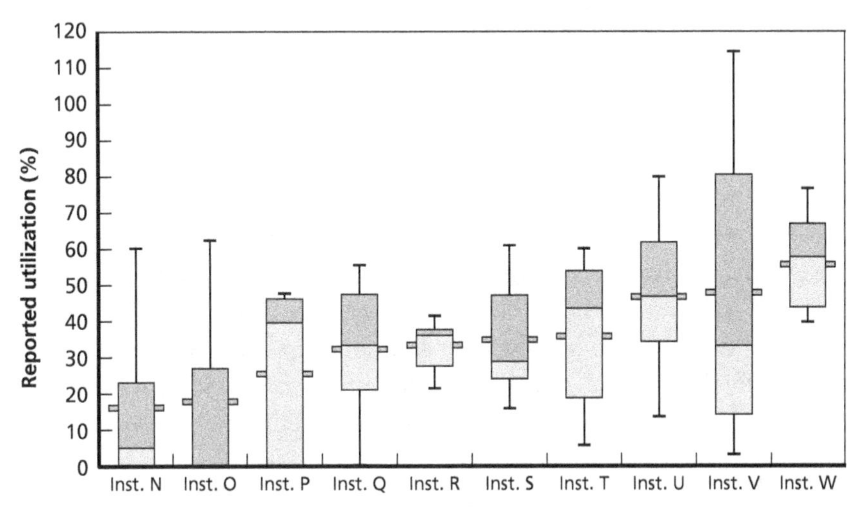

As with CCTT, there is a great deal of variability in usage across training sites and evidence of surges with high usage and periods of zero usage. And, again as with CCTT, although we know the number of company-level units that are resident at an AC post, we do not have good explanations for the variations in average usage or for the variation within sites.

Table 5.7 summarizes recent soldier throughput and system utilization metrics that PEO STRI and the TSAID collected for AVCATT.

As with CCTT, AVCATT the existing measures may overstate utilization. Although we do not have information about a typical day of usage for the AVCATT, the program does collect other information that points to this conclusion. The AVCATT program compares the cost of simulated flight hours to real flying hours (minus munitions expenditures) to show the cost trade-off between flying a simulator and flying a real aircraft. The costs per flying hour for real aircraft are calculated every year as part of the flying hour program. This comparison is conducted with a spreadsheet referred to as the "Blade Hour Comparison

Tool." It assumes that the facility is open for up to nine half-day training periods for ARNG locations and ten half-day training periods for AC locations, for an average of 9.5 periods per week. It also assumes 42 weeks of training availability for ARNG locations and 44 weeks for AC locations. Unavailable weeks are used for planned system maintenance and upgrades. Then, it considers how the cost per simulated flying-hour changes based on utilization assumptions, as summarized in Table 5.8.

GFT Utilization

The data available to understand utilization of GFT are different from those collected for CCTT and AVCATT. In contrast to CCTT and AVCATT programs, the GFT program lacks a centralized CLS contract that can provide insight into the daily operations of GFT sets. As a result, there are no contract requirements to provide data back to the program. In addition, GFT servers and systems are not designed to

Table 5.7
Aviation Combined Arms Tactical Trainer Utilization for Active and Reserve Locations, FYs 2014–2016

FY	Soldiers Days per Set per Year	Total Soldier Days per Year	Average Utilization (%)
2014	Unknown	Unknown	26.50
2015	83	7,615	22–26[a]
2016	83 ±19	76,02	38 (±5%)

[a] Data based on two utilization reports from PEO STRI and TSAID. Soldier days has been normalized to 23 sets. AVCATT FY 2016 data normalized to four quarters and a confidence interval added because reporting did not include the 4th quarter.

Table 5.8
Time Usage Assumptions in Blade Hour Comparison for Unit Aviation Combined Arms Tactical Trainer Training

	Aircraft Simulators Used Each Week	Simulated Flight Hours per Training Period	Maximum Simulated Flying Hours
Partial utilization	2.5	2	47,190
Full utilization	6	3 to 4	201,564

automatically collect utilization data. Before 2015, even manual documentation of GFT utilization was not collected centrally. Although GFT systems were designed to provide training for individuals and crews and collective training for a variety of proponent tasks, the utilization data that are collected do not distinguish these levels of training. CCTT and AVCATT are designed to provide and track collective training at the platoon and company levels. These systems certainly also provide valuable training of individual- and crew-level tasks, but use of the systems for these purposes is not tracked. Therefore, all three systems' metrics combine and conflate individual, crew, and collective training, which limits what we can conclude about utilization for collective training.

To estimate GFT utilization, we relied on utilization reporting for the second quarter of FY 2016. (We also had usage reporting data from FY 2015 and the first quarter of FY 2016, but we were concerned that these were not reliable and did not reflect all fielded locations.) In the second quarter of FY 2016, there were 21 AC locations reporting for 30 GFT sets and another 29 ARNG locations, each reporting for one set. This represents a little over 40 percent of reporting for sets and sites fielded by FY 2016. We would expect many of the sets fielded in FY 2016 to not be fully operational for the year because it is unlikely that they were fielded at the beginning of the year. Unlike CCTT, the contract does not require utilization reporting, and utilization data were not reported for many of the locations with GFT sets.

Figure 5.6 shows the range of utilization of GFT sets reported in the second quarter of FY 2016.[4] At this point, GFT had even greater utilization rate variability than its CCTT and AVCATT counterparts, but it was a relatively new system that was being implemented in sites during the course of this research. Thus, we do not display details of the variability in use.

---

[4] Note that utilization can be greater than 100 percent for a site if the sets were used for more hours than the time calculated as being available for the site. For the AC, available time typically consists of weekday business hours, and for most AC sites, available hours range from approximately 400 to 1,000 hours per quarter, with an average of approximately 680 hours. For most ARNG sites, there are 192 available hours per quarter. Available time includes scheduled downtime for maintenance.

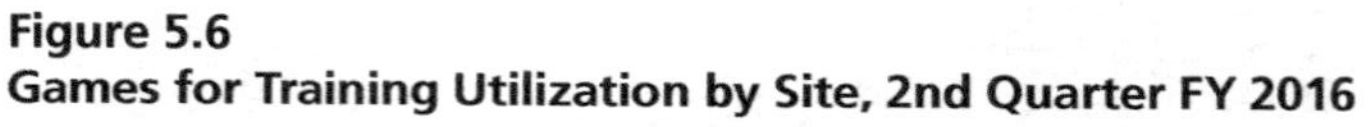

Figure 5.6
Games for Training Utilization by Site, 2nd Quarter FY 2016

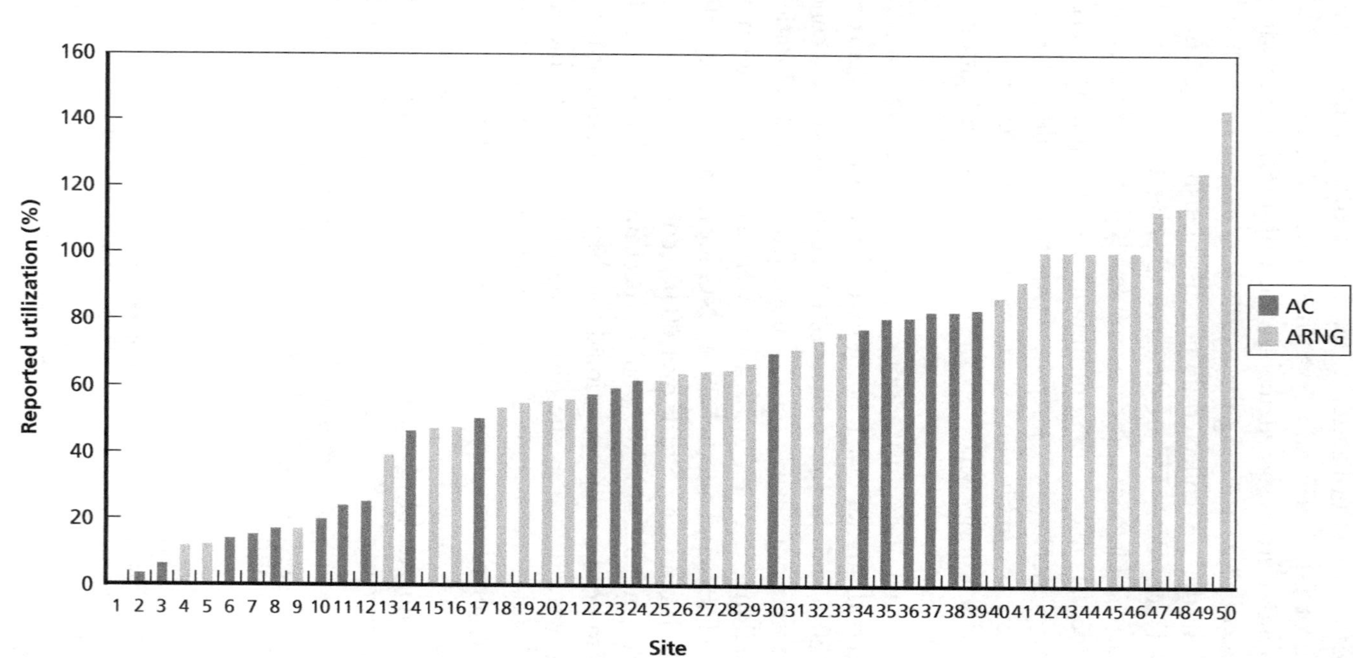

NOTE: The percentage of reported utilization was calculated by dividing the number of days the facility was reported as used by the number of days the facility was available for use.

Table 5.9 provides additional detail on number of soldier days per set per year and soldier days per year. As noted earlier, unlike AVCATT and CCTT, the GFT program is still in fielding and reporting is not mandated, so there is some uncertainty around utilization, but because the costs are for an average of 137 sets in the POM, we normalized usage reported to 137 sets in Table 5.9 for FYs 2015 and 2016. As the table shows, the number of soldier days reported increased from FY 2015 to FY 2016, even when normalized to the same number of sets, but the average utilization rate still remained near 40 percent ±8 percentage points, making it very similar to CCTT utilization reported in Table 5.5. Because we did not have data on which units were accessing the GFT sets for training, we could not provide a surrogate for the number of units that were in the potential pool of GFT users as we did for CCTT and AVCATT. Because the reporting is only for a sample of the GFT sites, we calculated a confidence interval for the average utilization

As with CCTT and AVCATT data on utilization, GFT data do not provide a clear picture of a typical day of usage. FY 2015 usage reporting for GFT does not indicate how much time the facility was reserved but does distinguish preparation and recovery time, exercise execution, AARs, and exercise teardown. FY 2016 usage reporting excludes preparation and recovery time. As shown in Table 5.10, the majority of hours in the GFT facility are devoted to exercise execution rather than troop-leading procedures. Troop-leading procedures are crucial, but as noted earlier, simulation equipment is not needed for

Table 5.9
Estimated Usage and Reported Games for Training Site Utilization, FY 2015 and Partial FY 2016

FY	Estimated Soldier Days per Set per Year	Estimated Soldier Days per Year	Average Reported Utilization (%)
2015	1,498	205,271	35
2016	1,965	269,173	48
Average	1,706	237,222	38 (±8%)

NOTE: FY 2016 data are limited to quarters 1 and 2. Both FY 2015 and FY 2016 do not include all fielded sites, so soldier days have been annualized to usage at 137 sets.

Table 5.10
Games for Training Utilization

FY	Exercise Execution	Exercise, AAR, and Exercise Teardown (%)	Exercise, AAR, and Preparation and Recovery Time (%)
2015[a]	66	78	100

[a] Reported.

planning and preparation; units can make a terrain map and conduct a rehearsal in any room with enough open floor space, rather than tie up a simulation facility. Ideally, the Army should work toward using the facility for the maximum value it can provide: exercise execution and AARs. This focused usage appears to be happening more with GFT facilities than with CCTT.

Cost-Effectiveness

By combining available cost and utilization data for CCTT, AVCATT, and GFT, we developed measures that compare their cost-effectiveness (see Table 5.11). The measures compare the systems in terms of cost per task, event, soldier day, and platoon hour. In general, as we suggested in Table 5.1, these measures indicate that CCTT and AVCATT are much less cost-effective than GFT. Because of limitations in the way that data are collected and reported, however, each measure cannot be estimated for all three training systems.

First, we calculated a cost-per-task metric. Given the tasks trained by each system in Table 2.5 (CAC-T, 2014), we know the total number of tasks that CCTT and GFT are capable of training for an ABCT. (AVCATT was not included in that study, so we are unable to compare the tasks in AVCATT and GFT.) Results indicate a total cost of approximately \$5 million per task per year for the CCTT modules and a cost of approximately \$1 million per task per year for GFT.[5] (If quality data were available, it would be worthwhile to analyze whether

[5] If the number of personnel for scenarios is higher or lower than anticipated, this GFT estimate could vary by ±20 percent but would still be lower than CCTT.

Table 5.11
Comparing Costs

	CCTT	AVCATT	GFT
Cost per ABCT task the system can train	$5 million	Unknown	$1 million
Cost per ABCT task trained per soldier per training day	$63	Unknown	$3
Cost per platoon training event[a]	$24,000	Unknown	Unknown
Cost per platoon direct training hour	$3,000	Unknown	Unknown
Cost per training event per soldier	$1,500		
Cost per potential soldier training day at current usage[b]	$750	$7,000	$200

NOTES: Costs estimates have been rounded. All costs are in dollars unless labeled as millions.

[a] A *training event* is defined as two days (four four-hour blocks) of facility use.

[b] These are also the costs per day shown in Table 5.1.

CCTT trains to the task five times better than GFT.) We then included the number of soldiers receiving training to calculate a cost per soldier per task.

Next, we used information from our CCTT site visit to explore the cost of training an ABCT in the CCTT. Our team's observations of a platoon training event and the typical costs of using the facility indicate that a soldier would typically spend two days at the facility. Typically, a platoon has 16 soldiers, so a two-day training event for 16 soldiers costs a little under $24,000. In an average training event, the platoon will receive between four to six hours of hands-on training time, at a cost of up to $3,000 per hour. The potential cost to train a soldier for one day is about $750 and for a two-day training event is about $1,500. These costs can be further deconstructed to a cost per task per soldier per training day of $63.[6] We were not able to observe a

---

[6] Prior research on GFT quantified the number of tasks trained by the system, shown in Table 2.4. These data, in combination with utilization reports, which reflect the number of soldiers trained, allow calculation of cost per task per soldier day.

typical AVCATT or GFT training event, so we were unable to calculate similar collective training event and hourly costs. We did not have the number of tasks for AVACATT and so did not calculate that here.

The final metric is the cost of a potential day in the facility per soldier. For this measure, we have estimates of the number of potential soldier training days for each program.[7] For CCTT, we divided the total cost of providing services ($64.62 million from Table 5.2) by the number of soldier days at the facilities (86,019 from Table 5.5), yielding an average cost of about $750. Using similar logic for AVCATT, we calculated a daily cost per soldier of about $7,000. For GFT, the cost per soldier day metric comes to about $200, which is significantly lower than CCTT and AVCATT. These results suggest that the quality of training with CCTT and AVCATT would need to be many times better or provide substantially greater throughput than GFT to be a better investment than GFT.

Summary

Our analyses show that the Army currently spends similar amounts annually on CCTT and AVCATT and less on GFT, as shown in Table 5.1. While we do not have information on the training effectiveness of each system, we can conclude that GFT is much less expensive than CCTT and AVCATT when the tasks trained are taken into account. Increasing utilization of CCTT and AVCATT will not make these systems less expensive than GFT. In addition, GFT has broad applicability to various forms of collective training beyond what AVCATT and CCTT can offer. We support efforts to collect improved utilization data that indicate how many unique soldiers receive training; distinguish how much of the training time involves direct interaction with the simulator and how much is for indirect activities, such as preparation for training and debriefing; and, for GFT, differentiate the types and levels of skills trained.

---

[7] It is possible that the full day is not being used for collective tasks but is also being used for individual- or crew-level task training, Therefore, we refer to the potential number of collective training days.

CHAPTER SIX

Summary, Recommendations, and Conclusions

As discussed in Chapter Three, user attitudes toward PSME are often more positive than warranted, given the empirical findings about effects of physical fidelity on training outcomes. Assumptions about the value of PSME for collective training may be reflected in the long-standing emphasis in the Army training community on training equipment relative to training design. Nearly 20 years ago, Salas et al. (1998) noted that, while there are clear specifications for technical features in development of simulators in aviation, there are few regulations and little guidance for the content and delivery of training using the simulators. The authors also pointed to the discrepancy between funding for equipment and funding to advance and apply knowledge of learning processes in SBT. They concluded that the

> challenge to training developers and simulator designers is to develop systems that use technology to promote learning. To achieve this goal, there will need to be a shift in focus from the designing of simulation for realism (and hope that learning occurs) to the design of human-centered training systems that support the acquisition of complex skills. (Salas et al., 1998, p. 199).

Stewart, Johnson, and Howse (2008) drew similar conclusions, stating that "a paradox exists in the aviation training community; that is, the technological base for simulation continues to evolve at a rapid pace, while the training programs supporting them have shown very little change over the past 20 years," and that "the scientific literature has demonstrated the primacy of proficiency-based training methodology over fidelity, yet the institutional bias in favor of fidelity persists" (Stewart, Johnson, and Howse, 2008, p. 16). Similarly, GAO analyses of virtual training in the Army show that these issues persist and that the Army devotes insufficient resources to comprehensively assessing the effectiveness of virtual devices, e.g., by conducting postfielding analyses. GAO also found that, while the Army has sought opportunities to increase the use of SBT, it has not developed the performance measures needed to determine the right mix of training approaches and has not obtained the cost information needed to assess the value of its training investments (GAO, 2013; GAO, 2016).

The imbalance in focus on training equipment relative to training design is likely one of the factors contributing to persistent assumptions about the need for PSME for collective training. As described in Chapter Three, these assumptions are well documented in the research literature, and they conflict with empirical findings on how physical fidelity affects training outcomes for individuals and teams. Similarly, while responses in interviews, focus groups, and surveys in this research acknowledged some value in GFT approaches for collective training, participants had much more positive attitudes about PSME simulators for collective training.

However, as discussed in Chapter Four, the interviews and focus groups revealed mixed views of both PSME and GFT approaches among Army stakeholder groups (developers, trainers, and consumers). Representatives of each group reported that CCTT and AVCATT are valuable and important for platoon- and company-level collective training, and survey respondents likewise reported that AVCATT is valuable for collective training. Nevertheless, this view was tempered by complaints that the physical fidelity of such systems lagged changes in actual military equipment platforms and by a reluctance to train on PSME devices that do not match standard operating equipment. Capability and training providers also raised concerns about the costs required to maintain concurrency of AVCATT and CCTT with fielded equipment. Attitudes toward GFT approaches were favorable among capability providers but more mixed among training providers and consumers. Findings from interviews and focus groups that experienced gamers were more enthusiastic about the use of GFT approaches for collective training suggest that attitudes toward VME systems may become more favorable as diffusion of the systems increases and as personnel become more familiar and proficient with system use. Key drivers of utilization of all collective SBT systems include command emphasis, policy or requirements for use, commander knowledge and skill in using the systems and related training resources, and accessibility or ease of use. These factors are integral to the recommendations we outline in the next section for the future of collective SBT systems.

As discussed in Chapter Five, the Army's PSME systems are both more expensive and less cost-effective than what we estimate for GFT. Although the Army currently spends similar amounts annually on CCTT and AVCATT and a smaller amount on the GFT program, GFT is less expensive than CCTT and AVCATT once utilization is taken into account. While the Army does not collect information about training effectiveness associated with each system, by combining available cost and utilization data, we developed measures that suggest their cost-effectiveness. Such measures compare the systems in terms of cost per task, event, soldier day, and platoon hour. These measures indicate that CCTT and AVCATT are much less cost-effective than GFT. Moreover, GFT has broader applicability to various forms of collective training beyond what AVCATT and CCTT can offer.

In summary, PSME simulators are costly (in large part because of features that provide high levels of physical fidelity), and their cost may not be justifiable in light of extant research findings about how physical fidelity affects learning outcomes, on Army utilization rates, and constraints on access to and ease of use these systems.

Recommendations

We propose two complementary sets of recommendations based on these results. The first set includes recommendations to increase utilization of SBT, foster improvement in SBT design and delivery, and improve training evaluation. The second set includes recommendations to conduct experiments to determine whether empirical evidence supports a transition from PSME systems (AVCATT and CCTT) to

VME systems, such as VBS3, for collective training and, if so, whether to begin this transition. The two sets of recommendations are independent in the sense that the Army could choose to implement one and not the other. But they are complementary in the sense that they would become synergistic if the Army created mechanisms to link them. For example, increased utilization of VME could provide information useful for prioritizing needed improvements to SBT—if the Army takes steps to ensure that VME utilization is appropriately measured and analyzed. Similarly, as SBT improves, utilization may increase—if the Army takes steps to ensure that commanders are made aware of and are convinced of the improvements.

We offer a total of six recommendations, which are described in more detail in the following subsections:

1. Revise training policy and strategy to encourage use of collective SBT systems and begin to transition to GFT and to incorporate best practices for training planning, delivery, and evaluation.[1]
2. Improve and standardize measures of performance in collective SBT.
3. Improve access to comprehensive TSPs to support SBT.
4. Improve and expand utilization data collection.
5. Conduct one or more experiments or demonstration projects to compare effects of CCTT and AVCATT with VME approaches (e.g., GFT) on learning from training and training transfer for collective tasks.
6. After implementing recommendation 5, evaluate courses of action for continued use of CCTT, AVCATT, and VME approaches for collective training based on relative demonstrated effectiveness and costs of these alternatives.

Revise Training Policy and Strategy

We recommend revising training policy and strategy to require or encourage use of collective SBT systems and begin to transition to

---

[1] We also recommend encouraging use of existing legacy systems while they are available and funded.

lower-cost systems, such GFT. As we have described, extant evidence indicates that SBT with high levels of physical fidelity does not produce training outcomes that are superior to those from lower-cost SBT equipment. In addition to cost, high physical fidelity systems also have significant drawbacks in terms of user access and ease, speed, and cost of updating. Reports from participants in our interviews, focus groups, and surveys, as well as in Seibert et al. (2012), indicate that changes to operational systems outpace changes to such training systems as AVCATT and CCTT, which can result in the potential for negative transfer of training.[2]

Despite these issues, many end users and other stakeholders in the training community believe that high physical fidelity systems are important for collective training—or conversely, that systems with low levels of physical fidelity are less effective—perceptions that are grounded in institutional biases and might be perpetuated by sunk costs, as discussed in Chapter Three. Thus, changes in the organizational culture are needed to support increased uptake of GFT or lower physical fidelity equipment for collective training.

Changes in culture may be achieved, in part, through command emphasis (e.g., Meredith et al., 2017). GAO (2016) found that use of specific virtual training devices was highest when training strategies prescribed use of the devices, and, as reported in Chapter Four, command emphasis was associated with high utilization of SBT systems. Similarly, the U.S. Air Force (USAF) Ready Aircrew Program tasking memoranda and Air Force instructions for flying operations require use of simulators for training on particular tasks (see, for example, Air Force Instruction 11-2F-22A, 2006). Beginning in 2002, use of SBT steadily rose in response to an emphasis on distributed mission operations and increased availability of SBT equipment (Chapman, 2006; Chapman and Colegrove, 2013). After an analysis of cost savings, USAF policy changes requiring minimum amounts of

---

[2] We recognize that SBT with high physical fidelity may be needed for training individual-level physical or motor skills, such as escaping from a Black Hawk helicopter that is upside down in water or getting out of the cupola of a Humvee gun truck as it is rolling over—such skills require rapid reactions with high tactile recognition and spatial awareness.

SBT (a minimum of three SBT sessions per month per pilot) led to a marked increase in simulator usage beginning in FY 2012. Figure 6.1 displays USAF data on the number of simulated tactical sorties that were allocated on applicable USAF simulators for FYs 2005–2016.

Ready Aircrew Program tasking memoranda and Air Force instructions for flying operations specify not only minimum amounts of SBT but standardization of tasks and skill progressions under differing conditions. In general, these requirements are set by the commands in collaboration with the relevant wings. In addition, doctrine is complemented by resources and expertise. In 2011, USAF stood up its Distributed Training Center, which coordinates changes in technology and doctrine, manages the resources required to carry out SBT for distributed mission operations, and facilitates the planning and execution

Figure 6.1
Training Simulator Use in Air Combat Command Before and After Policy Changes

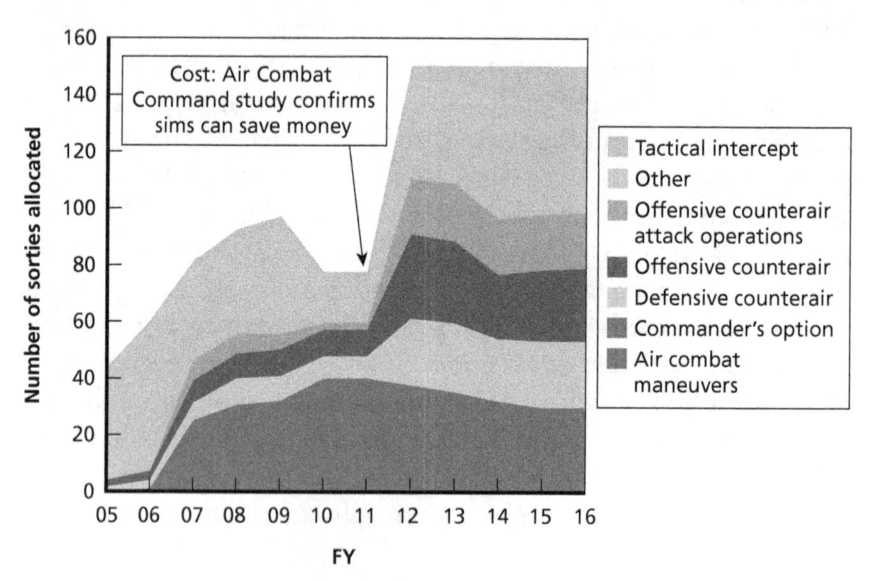

SOURCE: Ready Aircrew Program tasking data were gathered in support of a RAND Project AIR FORCE 2016 project, "Adversary Air Enterprise, Cost Effective Options for the U.S. Air Force."
NOTE: These data pertain to the F-15C, F-16C block 30/40, F-16C block 50, and F-22 airframes.

of complex distributed mission operation simulation scenarios. Over time, the culture of pilots and commands evolved to support greater use of simulated sorties. In sum, to enable greater uptake of SBT, the USAF developed detailed skill progressions and associated curriculum support (such as TSPs) that were accepted by trainees; stood up an organization to integrate, coordinate, and manage all aspects of SBT; and was able to require use of SBT to complement—and eventually replace—some elements of live training (Chapman, 2006; Chapman and Colegrove, 2013).

GAO's (2016) review of Army training regulations and other guidance—specifically, AR 350-38 (2013), TRADOC Pamphlet (TP) 350-70-13, AR 350-1 (2014), and AR 71-9 (2009)—found little in the way of requirements relevant to use of SBT devices, such as training time or target usage metrics. Only one document, TP 350-70-13 (2014), requires identification of specific tasks to be trained.[3] We agree with GAO's recommendations to update training policy to emphasize SBT and note that the most recent version of TP-525-8-2 (2017) puts substantial emphasis on the use of virtual, constructive, and gaming technologies in Army training.

In addition, we recommend that training policy and strategies require or encourage best practices with respect to training planning, delivery, and evaluation. GAO (2016) found that Army training strategies often did not specify how to accomplish or evaluate training tasks when using virtual training devices, even when strategies require use of these systems. GAO (2016) also reported that Army policies do not specify requirements for postfielding analyses of training effectiveness (and in GAO's case studies, the incidence of and approaches to conducting such analyses were highly variable). Revisions to policy and strategies for planning and delivery of training should be consistent with established learning principles (although research supports training to proficiency, rather than specifying training for a set amount of time; see, for example, Chapman and Colegrove, 2013; Stewart et al., 2008). For example, in describing development of a training strategy,

[3] These regulations are AR 350-38 (2013), TP 350-70-13 (2014), AR 350-1 (2014), and AR 71-9 (2009).

TP 350-70-13 (2014) should specify use of TSPs, which are roughly analogous to the USAF's specifications of standardized tasks and skill progressions.[4] We also recommend that the Army revise relevant policy documents to require (rather than encourage) training evaluation. TRADOC Regulation 350-70 (2017) specifies evaluation as part of the Army's instructional design framework (analysis, design, development, implementation, and evaluation), but training developers would benefit from more-specific guidance about evaluation methods.

If the Army were to adopt the recommendation to require or strongly encourage increased use of SBT, along with best practices for training planning, delivery, and evaluation, we expect that unit proficiency at collective tasks and unit readiness would benefit from the increased use of a training approach with established effectiveness. In turn, the Army as a whole would improve its return on the substantial investments needed to develop, operate, maintain, and upgrade SBT systems.

Improve and Standardize Measures of Performance in Collective SBT

We recommend using a combination of objective and subjective measures. Objective measures of performance can be obtained directly from SBT equipment, and measures of trainees' responses that might be associated with performance can be obtained from devices that monitor trainees' physiological responses. Subjective measures can be obtained from instructors or observer-controllers in the form of brief, structured assessments (e.g., scorecards or checklists) and from trainees themselves in terms of the value of SBT and their self-efficacy for relevant tasks.[5] Measures should link performance to training objectives and events, for example, as described in TP 350-70-1 (2012). Criteria for these measures could be derived from T&EOs associated with CATS. In addition to using performance measures to evaluate responses to discrete training events and provide input to AARs, they could be used to track performance over time or repetitions of an event

---

[4] The next two subsections address the content of training evaluations (recommendation 2) and the content and development of TSPs (recommendation 3).

[5] These measures can be used in steps 4 and 5 of the EBAT process described later.

to assess improvement (or conversely, skill decay), track performance across different events to assess more-complex skill development and adaptability (application of knowledge and skills to new or nonroutine situations), and assess training effectiveness at the program level by aggregating results across events and units.

Objective Measures

Automated Measures from Simulators

SBT equipment generally has the potential to provide automated, objective feedback about trainees' performance during training events. This is a goal of Objective Assessment of Training Proficiency (Objective-T), a reporting method the Army has adopted to provide more-objective measures of training readiness. Objective-T provides concrete criteria to determine readiness levels, such as demonstrating proficiency of mission essential tasks at greater than or equal to 90 percent, 80–89 percent, and so forth (the ranges may vary for different tasks and conditions). Each criterion range is associated with a rating of fully trained, trained, practiced, marginally practiced, or untrained (Headquarters, Department of the Army, 2017). The training events in which these assessments are carried out occur under conditions reflecting different operational environments, such as day versus night and single versus dynamic threats.

In addition to providing objective data, automated data capture eases evaluation burden for instructors or observer-controllers that can result in skipped or missing ratings (Dwyer et al., 1999; MacMillan et al., 2013). For example, the Air Force Research Laboratory's Warfighter Readiness Research Division developed the Performance Effectiveness Tracking System (known as PETS) for distributed simulation and live training environments (see Portrey, Keck, and Schreiber, 2006, for a review). These systems capture a wide range of objective air-to-air and air-to-ground combat performance measures in real time at the team, interteam, and teams-of-teams levels. Measures include outcomes (e.g., fratricides, mortalities, missiles fired that result in a kill), processes (e.g., time spent within the minimum abort range, point at which the aircraft can no longer turn to defeat a radar threat), and use of munitions (e.g., types of weapons shot by whom and at whom, altitude and loft angle when the pilot launches a missile). In addition, systems have been developed to capture objective performance data in networked SBT and process them rapidly to generate output (e.g., graphs, tables, still or animated views of trainees' actions). These data can be used to provide feedback to trainees in real time (e.g., at critical decision points during training events to scaffold training, i.e., to dynamically adapt training events to trainees' actions), in AARs, or to analyze performance after training events (e.g., Brown et al., 1997; Chen et al., 2007; Gately et al. 2005; Sadagic et al., 2013). Some systems not only provide feedback about trainee performance but can generate "what-if" scenarios or demonstrate the causal relationships between trainees' decisions and outcomes (e.g., Chen et al., 2007; Sadagic et al., 2013). Salas et al. (2009) also discuss the potential for automated systems that use simulator data to scaffold training.

Physiological Measures

The increasing prevalence of wearable devices presents many opportunities to monitor measures of training participants' physiological responses in real time (Sung, 2015), and physiological measures can be used to determine the appropriate level and/or dimension of fidelity. For example, Schnell et al. (2009) presented a system for aggregating electroencephalogram (EEG) and electrocardiogram (ECG), pulse-oximeter, respiratory, galvanic skin response, and vision tracking data to evaluate cognitive workload in real time. Tests of the system in a distributed training simulation environment have shown that EEG scores react relatively quickly to increased task demands, whereas changes in heart rate tend to lag from 30 seconds to 1 minute. The authors suggest that that these metrics could be used to automatically adjust training task demands to maximize learning. Other studies using physiological measures in SBT to assess cognitive workload and stress in military team tasks have found that patterns of physiological responses differ for novices and experts, which has implications for fidelity in training design (e.g., Stevens et al., 2013).

Other work by Schnell and his colleagues (Schnell, Postnikov, and Hamel, 2011; Schnell et al., 2012) and by Klyde et al. (2013) find that changes in fidelity of simulators, or comparisons between simulators and real flight, are associated with changes in physiological measures indicative of cognitive workload. However, differences in results from these studies suggest the need for additional research to identify the association between specific physiological cues and types and levels of fidelity.[6]

Most research to date has used physiological measures to assess workload, but there is increasing research on the use of such cues to assess task performance (Galán and Beal, 2012; Guru et al., 2015; Sciarini et al., 2014) for individuals and teams. These studies typically examine the associations of patterns or sequences of physiological measures, such as EEG and ECG, with measures of team processes (e.g., patterns of communication) and one or more measures of task performance, including objective metrics, observer ratings, and trainee self-reports (e.g., Johnson et al., 2013; Stikic et al., 2014; Stevens et al., 2013). Additional research on the use of physiological measures to evaluate performance in military team and collective training tasks may prove fruitful for both training design and performance assessment.

Subjective Measures

A number of behaviors relevant to collective performance, such as team communication and coordination, cannot be scored efficiently in an automated fashion. For these behaviors, instructors or observer-controllers can provide ratings using scorecards on paper or on hand-

[6] Schnell and colleagues (Schnell, Postnikov, and Hamel, 2011; Schnell et al., 2012) found that increases in fidelity, in terms of the number of visual channels, led to a reduction in heart rate (e.g., Schnell et al., 2011; Schnell et al. 2012). The authors suggest that more channels corresponded with a lower cognitive workload because maneuvering tasks become easier with additional visual information. However, heart rate increased during real flight, arguably because of changes in other cues, such as motion and sound, as well as the knowledge that one is operating a real and potentially dangerous airplane. Klyde et al. (2013) found that physiological measures (EEG, ECG, and eye blinking patterns) were sensitive to differences in levels of functional fidelity in simulations and between high fidelity simulators and actual flight. Higher fidelity simulations resulted in a slightly elevated heartrate and workload, as derived from physiological measures. In contrast with the findings of Schnell, Postnikov, and Hamel (2011) and Schnell et al. (2012), Klyde et al. (2013) found that cognitive workload in real flight did not increase significantly over high fidelity simulators. These results suggest the need for additional research to identify the association between specific physiological cues and types and levels of fidelity.

held digital devices. Using digital devices can facilitate rapid analysis and integration for AARs and for ongoing evaluation, as described later in this chapter. Ratings can be conducted in real time or by observing recordings of training events.

Numerous scorecards have been developed to evaluate performance in military training (for reviews, see Dwyer and Salas, 2000; Lawson, Kelley, and Athy, 2012; and Rienerman-Jones et al., 2015). Some scorecards consist of checklists, where raters assess whether particular actions occurred or not, and others consist of rating scales, in which raters judge the quality of responses. Most of these instruments are customized to particular events. Table 6.1 presents examples of scorecards for training evaluation. Table 6.2 presents an example drawn from one of these checklists.

Checklists differ from the simple "go, no-go" ratings typical of Army training by the type and granularity of the tasks being rated. For example, the criterion "gained and/or maintained situational awareness," which is the first go, no-go performance measure of company and platoon CATS for "movement to contact," is quite broad and may consist of many subtasks or behaviors. A checklist consisting of more-specific behavioral indicators associated with attaining situational awareness will produce more-verifiable and -consistent ratings and will be more informative to trainees, helping them diagnose problems and identify solutions. Likewise, use of detailed, structured scorecards can provide concrete feedback to trainees and limit subjectivity in judgments, consistent with the goals of Objective-T.

When using scorecards, there is a trade-off between comprehensiveness of the feedback and the cognitive load for raters. Some scorecards have used large numbers of items, which is burdensome to raters and can result in a large number of missing ratings (e.g., Dwyer et al., 1999; MacMillan et al., 2013). MacMillan et al. (2013) found this to be the case with an instrument consisting of 26 items; the other measures discussed in Table 6.1 contain many more questions. Rater burden may be particularly problematic when rating collective events in distributed environments, given the number of trainees, complexity of training tasks, and geographic distribution among participants (Dwyer et al., 1999). Alternatives to reducing the number of items

Table 6.1
Examples of Scorecards to Rate Performance in Training

Instrument	Sources	Purpose	Type	Description	Notes
Targeted Acceptable Responses to Generated Events or Tasks ("TARGETs")	Fowlkes et al., 1992; Fowlkes et al., 1994	Naval aircrew coordination	Checklist	Raters evaluate expected behavioral responses to training events as either "hits" or "misses." The number of items depends on the event. Raters do not need to be SMEs because the specificity of the instrument does not require expert judgment. See Table 6.2 for an example of items within a flight segment.	• Ratings can be aggregated by segments, events, or categories of behaviors (e.g., backup behaviors that are expected across events or segments) • Studies using this instrument (Dwyer et al., 1999; Fowlkes et al., 1992; Fowlkes et al., 1994) have found that it has high interrater reliability (consistency in ratings from different observer-controllers), sensitivity (distinguishing better or worse performance), and face validity (perceived as relevant by users)
Distributed Mission Operations Gradesheet	Krusmark, Schreiber, and Bennett, 2004	F-16 air combat performance	Rating scale	SMEs rate 40 items using six-point scales; examples of items are "postmerge maneuvering," "mutual support," and "visual lookout." Response options are N/A = not applicable; D = dangerous; and five additional options ranging from O = "performance indicates a lack of ability or knowledge" to 5 = "performance reflects an unusually high degree of ability."	• Implemented on handheld devices • Ratings captured improvement in performance as training progressed but had poor interrater reliability, lacked sensitivity, and did not distinguish among performance indicators

Table 6.1—Continued

Instrument	Sources	Purpose	Type	Description	Notes
Scenario-Based Performance Observation Tool for Learning in Team Environments	MacMillan et al., 2013	F-16 air combat performance	Rating scale	SMEs rate performance on 26 items tied to scenario periods, rather than events. Items are rated on five-point, behaviorally anchored scales. For example, response options for an item assessing mutual support in the team include 1 = "no mutual support"; 3 = "detached mutual support"; and 5 = "visual mutual support."	• Implemented on handheld devices • Initial findings of MacMillan et al., 2013, suggest that ratings are reliable and sensitive • This article illustrates the labor-intensive, iterative processes involved in instrument development
Nontechnical military skills (B) ("NOTEMILs")	Tsifetakis and Kontogiannis, 2017	F-16 individual and team skills	Rating scale	Raters evaluate situational markers (descriptions of poor and good performance) on items reflecting four groups of skills: situational assessment (19 items), managerial and leader skills (17 items), team cooperation (17 items), and decisionmaking (11 items). Items are rated on seven-point scales reflecting levels of expertise as described by Dreyfus and Dreyfus, 1986 (compare Tsifetakis and Kontogiannis, 2017): novice, experienced beginner, practitioner, knowledgeable practitioner, expert, virtuoso, and maestro.	• Tsifetakis and Kontogiannis, 2017, found that nontechnical military skills (B) ratings predicted mission essential competencies (tactical and technical skills) as rated by an independent SME, although predictions varied for nontechnical military skills skill groups and technical tasks • This study also showed good reliability (interrater agreement) among trained raters

Table 6.2
Example Section of Targeted Acceptable Responses to Generated Events or Tasks Checklist

Segment	Event	Targeted Behaviors	Hit
Prior to liftoff	Ship's tower provides erroneous weather during takeoff clearance	• Pilot flying questions weather information	
	Takeoff clearance given	• Pilot flying acknowledges takeoff clearance	
		• Takeoff checklist completed using challenge-and-reply method	
		• Pilot flying asks aircrewman if cabin ready for lift	
		• Pilot flying alerts crew that he/she is taking off	

SOURCE: Fowlkes et al., 1992.

include using multiple raters (e.g., assigning raters to observe different aspects of the event) and conducting or completing ratings after the event using video recordings (MacMillan et al., 2013).

Whereas the measures described above are context-specific and, in some cases, are tied to specific triggers or events in training, the Anti-Air Teamwork Observation Measure (ATOM) assesses team processes relevant to a wide range of situations (Johnston et al., 1997). ATOM was developed for the TADMUS program, sponsored by the Office of Naval Research. This rating scale consists of 11 behavioral items corresponding to four dimensions of team processes: information exchange, communication, supporting behavior, and leadership and initiative (Smith-Jentsch, Johnston, and Payne, 1998). For example, the "error-correction" item reflects supporting behavior (see Figure 6.2). Studies using ATOM provide evidence for its internal consistency reliability, construct validity, convergent validity, discriminant validity, and diagnosticity for assessing team training interventions (see Smith-Jentsch, Johnston, and Payne, 1998).

Figure 6.2
Example Item from the Anti-Air Teamwork Observation Measure

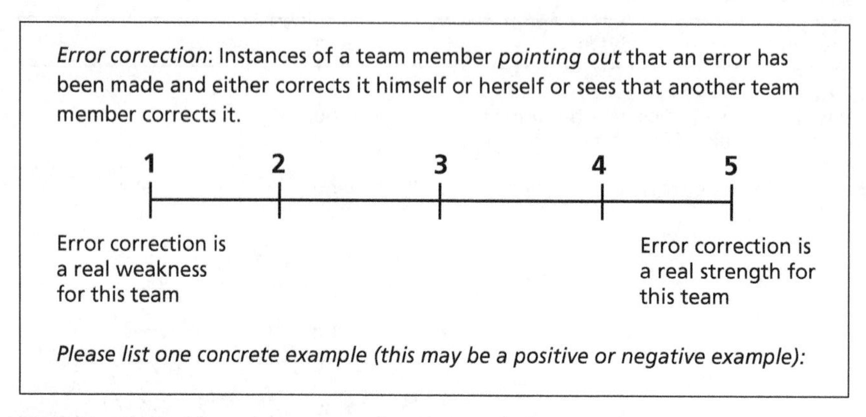

SOURCE: Adapted from Johnston et al., 1997. Used with permission from J. Johnston.

Integration and Reporting

The amount of data that automated methods can produce, in conjunction with ratings from human judges, requires integration so that instructors can use it and so that it provides useful feedback to trainees. Providing AAR templates or automated or semiautomated AARs populated with training-session data would reduce the burden on the company commander and, potentially, improve the quality of the feedback to trainees in SBT.[7] Several authors describe systems that integrate and analyze multiple data sources. Chen et al. (2007) review a number of such systems and describe the After Action Intelligent Review System developed for the Marine Corps' Combined Arms Command and Control Trainer Upgrade System, which is a live, virtual, and constructive collective training environment. The review system combines a variety of simulation and communications data and tools, including use of speech recognition capabilities to extract content from radio communications, for use during AARs and in subsequent analysis. Theoreti-

---

[7] Many multiplayer first-person-shooter games, such as "Call of Duty," have incorporated automated leaderboards that track players and display player achievements, such as numbers of kills, deaths, ratios, headshots, and other attributes of play. Some also have automated highlight videos to replay where critical events occurred in the game, such as a player character's death.

cal frameworks relevant to performance assessment should be used to guide development of such tools (Endsley, 2000; Fowlkes et al., 2005).

Provide Access to Comprehensive TSPs

Our third recommendation is to ensure that commanders have access to comprehensive TSPs to facilitate planning, delivery, and assessment of SBT and are encouraged to use these resources to support training. Seibert et al. (2012) also identified the need for TSPs to support collective SBT. Access to TSPs will not guarantee that they will be used or used effectively to support training. We recommend training unit personnel in the use of TSPs. Professional military education can provide initial training and instruction on principles of learning, training, and training evaluation. Refresher training can occur through mobile training teams, web-based train-the-trainer modules (Johnston et al., 2016), and online user forums (Hallmark and Gayton, 2011), among others.

One of the findings that emerged from our interviews is that appropriate preparation for training is time consuming. As a result, some company commanders forgo virtual training or schedule virtual training but put little effort into planning or preparation. As described in Chapter Four, stakeholders reported that, even though TSPs for CCTT training are available, company commanders typically do not use them to design or evaluate training. Instead, company commanders prefer to use "free play" (see also Dwyer and Salas, 2000) or to specify the training task to Mission Training Complex staff (e.g., practice "hasty defense" tasks) and have contractors set up and run a scenario.

Likewise, as discussed in Chapter Four, some units put little preparation or effort into systematic assessment and recording of key areas for sustainment or improvement for discussion in AARs. Evaluation of unit performance tends to be informal and not necessarily tied to CATS performance measures. In addition, personnel who are responsible for training may not be well versed in relevant educational principles. For example, Mastiglio et al. (2011) found that, while instructors or observer-controllers have guidelines for conducting AARs, programs of instruction in formal Army training do not address the principles underlying effective AARs; instead, personnel tend to learn how to conduct AARs by observing others on the job. Therefore, we also recommend that the tasks in the contracts for support personnel include development of training plans and scenarios and facilitation of AARs.

TCM V&G has developed more than 90 TSPs to support training CATS in GFT. Company leadership should be made aware of the availability of these TSPs and the benefits of their use. In addition, we recommend assessing the quality of TSPs, which was beyond the scope of this effort, and ensuring that TSPs are developed or adapted following a process, such as EBAT, as discussed in Chapter Three. EBAT, which has been researched in several studies in military training environments (e.g., Oser et al., 1999; Smith-Jentsch, Johnston, and Payne, 1998), involves the following steps (Dwyer et al., 1999):

1. Identify training objectives: critical tasks, conditions under which tasks are performed, and standards for task performance.
2. For each training objective, identify learning objectives specifying critical behaviors associated with each training objective.
3. Create scenarios or scripts consisting of events or triggers that provide opportunities for trainees to perform the critical behaviors and to practice tasks with increasing degrees of difficulty. Events and triggers can be used to stimulate routine or nonroutine situations, technical performance, and team interactions and to inject increasing degrees of task difficulty. Scenarios do not necessarily need to replicate real-world events but instead need to prompt the required behaviors associated with critical tasks (Kozlowski and DeShon, 2004).
4. Develop measures and instruments to assess performance in training, that is, responses to events and triggers.
5. Following training, analyze performance measures and use results to provide feedback about training and learning objectives to trainees (typically in an AAR).

Development of CATS—e.g., as described in TP 350-70-1 (2012)—provides the products needed for steps 1 and 2 in the EBAT process. Step 3, providing scenarios with increasing levels of difficulty, is consistent with criteria of Objective-T to move trainees from a rating of untrained to trained. An example for armor platoons would be conducting a "movement to contact" in a simple operating environment with flat terrain, during the day, in clear weather, and against an inexperienced opponent, then working through a progression of difficulty to an environment with challenging terrain, at night, in the rain, and against a skilled opponent.

Given the potentially large number of critical collective tasks and events, one approach to identifying where to start would be to use one of the training design models or decisions aids described in Chapter Three. The Task and Training Requirements Analysis Methodology (TTRAM) model (Swezey et al., 1998) may be particularly useful in light of its focus on collective tasks and use of networked simulations for training and its foundations in industrial and organizational psychology and education principles. Figure 6.3 summarizes the TTRAM process. In brief, the model entails computing a skill-decay index based on task difficulty, degree of prior learning, and frequency of performance and a practice-effectiveness index based on the amount, fre-

Figure 6.3
Summary of the Task and Training Requirements Analysis Methodology Process

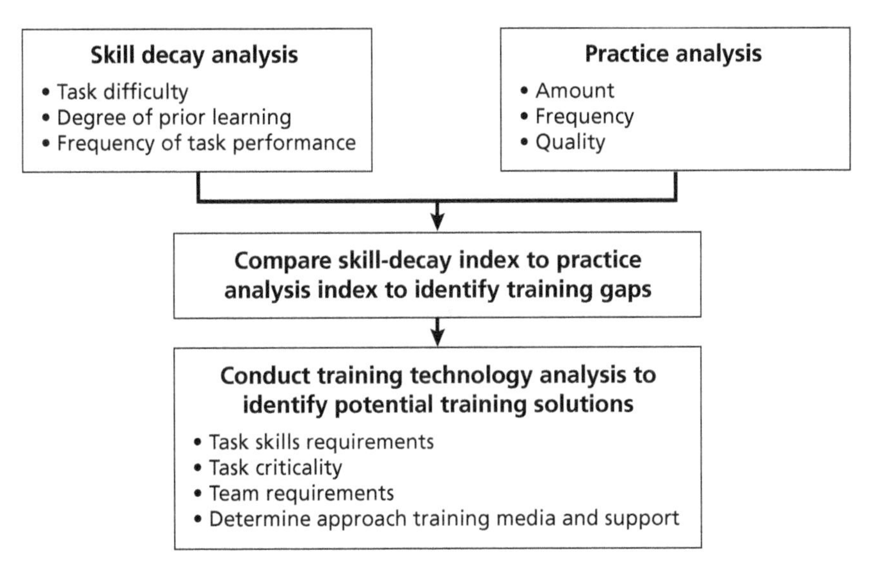

SOURCE: Adapted from Swezey et al., 1998.

quency, and quality of task practice. A comparison of these indexes may reveal training gaps. Tasks with larger gaps may be a good starting point for development of TSPs.

If the Army were to adopt the recommendation to make comprehensive TSPs accessible and encourage their use, we expect that it would improve training delivery and evaluation. We also expect that it will reduce time required for company commanders to plan training, which GAO (2016) identified as a factor that influences use of collective SBT systems. Adoption of this recommendation may not, however, affect the time available to units to participate in training.

Improve and Expand Utilization Data Collection to Evaluate Effectiveness of SBT at the Program Level and Support Research on SBT

In addition to the performance data described above, we recommend improvements to collecting system utilization data to assess program effectiveness. As described in Chapter Five, utilization is not measured consistently across CCTT, AVCATT, and GFT. Utilization rates for CCTT appear to overestimate direct training time on the equipment. We were not provided information on how AVCATT utilization rates are calculated.

GFT utilization rates provide a clearer breakdown of direct training time, but there was no centralized requirement to report usage data to the program at the time of this research; the systems are not instrumented to collect usage data automatically and therefore require the program to rely on manual reporting; and the reliability of data collected prior to FY 2015 was particularly open to question. And, as discussed in Chapter Four, utilization rates for these systems do not account for the number of units at a site that may have been deployed during the periods measured and, hence, were not available to train on the systems. Our analyses presented the average number of units at a location that could have been present during the period and did not appear to show a relationship between usage and number of potential user units. Future utilization reports should include the number of resident potential user units for each reporting period to better understand actual usage rates.

The Army Training Support Center is working to improve standardization and aggregation for all three systems. In particular, we recommend automatically capturing utilization when possible, which can reduce burden on company commanders, increases data reliability and validity, and facilitates data integration for analysis of utilization and costs at the program level. In addition, utilization data that distinguish how equipment and facilities are used (preparation time, exercise execution, AARs, and teardown) can enhance the accuracy of cost-effectiveness analyses. Finally, current reporting calculates soldier days but does not show how many unique soldiers received training. Reporting both numbers of unique soldiers and soldier days will provide insight into how much variation in training time individual soldiers might receive. Providing an automated method for tracking the participation and performance of individual personnel engaging in collective SBT could facilitate tracking of system usage and, perhaps more important, could facilitate analysis of how individual skills affect collective skills and vice versa.

It is also critical that data collection and aggregation for all future Army training system acquisitions must be explicitly designed for evaluation of training effectiveness. We acknowledge that improving training readiness is the primary goal of such systems but stress that being able to measure training effectiveness—the amount of skills and knowledge the system helps soldiers gain, as measured before and after training and over time—is key to improving the value of the systems to the Army and demonstrating that value as a return on investment in training technologies (in addition to providing trainees feedback about their success in training). As mentioned in Chapter Five, data on effectiveness and/or throughput are needed to justify the cost disparity between CCTT/AVCATT and GFT. Hence, contracting for systems should explicitly include the types of data valuable for such evaluations and clearly specify that all data be owned by, and easily accessible to, the Army's analysis communities.

To assess program effectiveness, the Army should consider using utility analysis, which estimates the economic value of personnel practices (e.g., Cascio, 1989; Schmidt, Hunter, and Pearlman, 1982). Utility analysis has been used primarily for employee selection but also has been applied to training and development programs (see Arthur et al., 2003; Cascio, 1989). Collecting more-accurate utilization data will provide inputs needed to assess the costs of training, which is one component of utility analysis. Assessing the benefits of training is much more challenging, however, because it requires estimating performance improvement associated with training and the dollar value of such improvement. Thus, consistent measurement and reporting of the performance measures described earlier (automated performance measures from simulators, SME ratings using scorecards) would be needed. According to GAO (2016), the Army is engaged in a pilot program to collect unit training activities, associated costs, and resulting training readiness that could be used for this purpose.

Adopting the recommendation to improve and expand data collection would offer the Army improved insight into both the effectiveness and cost-effectiveness of its SBT programs and provide data that could be used to support the improvement of SBT systems and support elements, including TSPs.

Conduct One or More Experiments or Demonstration Projects

We recommend that the Army conduct one or more experiments or demonstration projects to compare PSME (CCTT, AVCATT) and VME (e.g., VBS3) directly on learning from training and training transfer for collective tasks. As discussed in Chapter Three, although decades of research indicate that training for collective tasks does not require high levels of physical fidelity, definitive studies have not been conducted comparing PSME and VME. The studies we propose would create Army-specific evidence showing how SBT, in combination with effective training design and delivery, can support collective training. They could help corroborate previous findings suggesting (1) that if training is designed and supported appropriately, shifting from a reliance on PSME to VME will significantly decrease simulator life-cycle costs (i.e., for acquisition, operation, maintenance, and retirement) without decrements in learning and transfer of training and (2) that, compared to PSME training systems, VME can provide more opportunities for practice, which is essential for skill mastery and retention. That is, specialized expertise is not necessarily required to set up and run VME for training, which can allow more hours of access to the equipment per day and more days per week than current allocations.

The projects we propose to develop Army-specific evidence could range from simple pilots to randomized, controlled experiments. In addition to providing evidence about fidelity for training collective tasks, conducting one or more experiments or demonstration projects can build champions for the technology, which is important, given many stakeholders' beliefs about the value of PSME coupled with more variable views about VME approaches. Experiments or demonstration projects can also provide information about implementation, outcome measures, and policy to facilitate the transition to VME. There are trade-offs among study approaches in terms of these benefits, the scientific evidence produced, and the costs (see Table 6.3). For example, experiments are more complex and costly to conduct but provide much more information to guide future directions for SBT.

In all cases, the proposed projects would entail removing access to CCTT or AVCATT in sites that already have GFT sets (or replacing CCTT and/or AVCATT with new GFT sets), providing access to comprehensive TSPs (as discussed in the next section), and collecting a variety of measures:

- surveys conducted at the start of the intervention period and after some set time (e.g., six months) to assess participants' attitudes

Table 6.3
Benefits of Project Approaches

	Benefits				
Approach	Provide Scientific Evidence	Develop Champions	Gather Knowledge of Implementation	Inform Policy	Improve Usage, Cost, and Effectiveness Data
Experiment	+++	++	+++	+++	+++
Replicated cases	+	++	++	++	++
Simple pilot		+++	+	+	+

NOTE: "+" indicates additional benefits.

toward simulators and background characteristics; the survey developed for this study, reported in Chapter Four, could be adapted for this purpose
- system logs documenting VME system usage before and during the intervention period
- training performance during the intervention period using automated data capture and scorecards as described above; the experimental methods would include measuring subsequent performance in field training exercises at home station
- SME ratings of live or recorded AARs conducted during the intervention period.

Data would be analyzed to examine changes in user attitudes, system usage rates, and training performance from the beginning to the end of the project.

A field experiment would include (1) intervention groups, which would have GFT sets and TSPs instead of CCTT or AVCATT and would receive relevant education and training support, and (2) matched control sites, which would still have access to CCTT or AVCATT ("business as usual"). Intervention and control sites might be selected based on a history of high or low utilization of CCTT or AVCATT. Including a control group provides the strongest evidence that results in the intervention group occurred as a result of the intervention and are not due to other factors. The experiment would compare intervention and control groups in terms of user attitudes; system usage rates; cognitive load during training; performance in training; and transfer of training, i.e., performance in field training exercises and live fire events at the platoon and company levels, followed by performance in a combined arms live fire exercise. Experiments could also use physiological measures, such as EEG and ECG data (as described earlier), to assess the degree to which training in CCTT/AVCATT and in VME induces cognitive load.

While controlled experiments provide the strongest evidence on fidelity for collective training, carrying out such studies is challenging. First, a large sample of companies or platoons is needed to detect differences between experimental conditions or, if there are no differences found, to be confident that the results are meaningful. Second, SMEs must be available and willing to serve as raters and need to be trained to provide reliable and valid ratings. Third, awareness of the experimental conditions on the part of SMEs and study participants (e.g., platoon members) can influence results, and it can be difficult to keep these stakeholders naïve to conditions, particularly in a field experiment that occurs over a long period. Finally, if the experimental conditions have differential effects, this can influence readiness. If, for example, groups using VME show poorer performance, these participants would need additional training to make up for this decrement.

Replicated case studies could be similar to the experimental intervention described earlier, replacing CCTT or AVCATT with VME and TSPs at a number of sites that vary in factors that might affect uptake or effectiveness. This approach would not include control groups, making it less costly but also making it more difficult to draw conclusions about effects of the intervention.

A *simple pilot* might consist of conducting the demonstration project in a convenience sample at one or more sites, particularly in sites that may be the most receptive to a demonstration. For example, an infantry brigade combat team is being reconstituted as an ABCT at Fort Stewart; this change might provide a natural setting for initiating an approach to SBT, i.e., a "Live/GFT Training brigade" to test the impact of SBT without access to PSME (CCTT and AVCATT). Alternatively, sites with low CCTT or AVCATT utilization rates might be appropriate for a pilot. A pilot is the least costly of the proposed approaches but also provides limited evidence on which to base broader implementation.

Experiments or demonstration projects could be used to assess a range of other topics related to the adoption of VME and the support of improved TSPs. For example, studies could examine the effects of encouraging or requiring use of VME and TSPs (versus no encouragement or requirements) on how commanders execute training, utilization rates for SBT systems and TSPs; amount of practice attained; and outcomes, such as the speed, quality, and transfer of collective skills acquired. Studies could also examine effects of variations in the mix and order of VME and PSME to train collective skills.

Adopting the recommendation to conduct one or more experiments or demonstrations would likely produce compelling evidence that could be disseminated to improve Army commanders' understanding of the advantages of VME. According to GAO (2016), improved understanding can, in turn, contribute to increased use.

Evaluate Courses of Action for Continued Use of CCTT, AVCATT, and VME Approaches for Collective Training

The experiments or demonstration projects from recommendation 5 could provide evidence of greater, lesser, or equal collective training effectiveness of PSME versus VME for armor or aviation units. If, in contrast to what relevant literature suggests, there is evidence of greater training effectiveness for SBT using PSME (CCTT and AVCATT) over using VME, then the Army should analyze the trade-offs of using the more-costly technologies. For example, is the additional training value of PSME worth the differences in annual costs (e.g., equipment, maintenance, support, and upgrades)?

If the experiments or demonstration projects show that VME is as good as or better than PSME for collective training, we recommend that the Army begin to phase out use of CCTT and AVCATT for collective training. A plan for phasing out CCTT and AVCATT might begin at sites with low utilization, consolidating systems and placing them in warm storage while increasing the sites' GFT capabilities by providing hardware and personnel or providing GFT sets, if they are not already available. Depending on the length of the transition process, resourcing for CCTT or AVCATT at sites that still have access to these simulators should be aligned with anticipated utilization rates. A transition from CCTT and AVCATT to VME should be accompanied by training and education that addresses both the value of VME for collective training and instruction on how to use the system and supporting TSPs to plan and implement training.

Currently, use of CCTT, AVCATT, and VBS3 requires dedicated expert assistance of varying amounts to provide training. In general, however, the nature of the COTS hardware for delivery of VBS3-based or other VME-based training suggests that there is less need for hardware and network support personnel. The Army should aggressively seek to provide more soldier-enabled training via VME by making the GFT sets more turnkey for users. This could produce more hours of access to GFT facilities at lower support costs. To decrease the need for dedicated training support personnel, the Army should support development and use of user-friendly tools and software development kits that soldiers can use to build or tailor scenarios, terrain, opposing forces, weather conditions, and so forth. Such tools are common and widely used in gaming communities. Thus, we recommend exploring the feasibility of training unit personnel to prepare and execute training using VBS3 or future GFT software for collective training. To further support units' use of SBT, the Army should consider introducing an additional skill identifier to provide specific expertise in battalions, similar to tank master gunner and mission command digital master gunners. Soldiers serving in this role, which might be called "simulation master gunner" for SBT, would be trained to provide support and guidance to battalion and company leaders on how to create and tailor scenarios; organize training; and, generally, how to most effectively leverage SBT to increase unit training readiness.

If CCTT and AVCATT could be replaced with less costly and more accessible options, the Army could achieve several benefits. It could end continued investment in a training modality that is not state of the art, free resources that could be invested in a newer modality (VME) that is less expensive to operate and upgrade and for which there is evidence that it may be equally or more effective, and increase opportunities for unit training. In addition, along a longer timeline, the Army will be investing in new armored ground vehicles and, eventually, phasing out the M1 and M2 platforms. This change would make the manned modules of CCTT and MCCTT obsolete. So, instead of continuing to upgrade these PSME systems, the Army would have VME that could simulate new equipment and could even use that VME (and perhaps hybrids with PSME interfaces) to aid in the development of the new systems called *early synthetic prototyping* methods. Experiments using VME for such purposes are already being carried out at the USC Institute for Creative Technology (USC Institute for Creative Technologies, undated).

Research Strengths, Limitations, and Directions for Future Research

This research has a number of strengths. We used a wide range of methods and sources as a basis for our analyses and recommendations. We conducted a broad review of the literature to understand the landscape of research methods and findings pertaining to fidelity and training outcomes for collective tasks. We used interviews, focus groups, and surveys to understand experiences with and attitudes toward collective SBT from diverse stakeholders, including consumers, designers, and developers of training. Finally, we obtained data from archival sources to analyze utilization and costs of collective SBT. Observations of a company training exercise using CCTT and a mission training session for a pair of AH-64 crews in AVCATT, along with discussions with the contractors providing the training, also contributed to an understanding of how equipment for SBT gets used in practice.

The research also has limitations. The primary limitation is a function of the data the Army collects or would share with respect to system utilization and effectiveness. As described in Chapters Five and Six, problems with archival data—such as a lack of centralized data collection and management; ambiguity or inconsistencies in operationalizing use of SBT systems; and failure to measure outcomes, such as the number of soldiers trained or training effectiveness—and a lack of cost data from the sponsor limited the precision of estimates of utilization and costs. These issues are not limitations of our methods; in fact, they can be considered findings because they point to the types of data the Army needs to collect to evaluate program effectiveness. Centralized data collection and management, clearer definitions of system usage, and collection of a broader range of measures—particularly performance in training—will enable the Army to conduct more-rigorous program evaluations to support selection, use, and management of virtual technologies for collective training.

An additional limitation of this research centers on the representativeness of the data collected in the interviews, focus groups, and survey. Interviews and focus groups with larger samples of brigade staff and unit soldiers who are the consumers of training and from more locations would enhance the external validity or generalizability of the findings. In the case of the survey, higher response rates from USAACE and getting traction from personnel at MCoE, along with recruiting prospective participants with a broader range of Army experience, would strengthen the findings. The survey is available for the Army's use, and we recommend administering it on a regular basis to monitor attitudes toward and experiences with collective SBT technologies over time.

APPENDIX A

Examples of Platoon- and Company-Level Armor Unit Structures and Collective Tasks Taught in Collective Training

This research focused on different technologies that could support the training of collective tasks. The first assumption about teaching collective skills at the platoon or company level is that each level below the company—at the crew or platoon level—has achieved proficiency at that level. This is shown in Figure A.1 by boxes around sets of tanks; the boxes indicate that the battalion has already certified the tank crews and platoons below the company level as qualified before they engage in company-level collective training. This qualification includes gunnery and maneuver assessments.

The CATS for platoons includes 33 tasks; that for companies includes an additional 14 tasks. These collective tasks are listed in Table A.1.

Figure A.1
Structure and Examples of Tasks for an Armor Company and Platoons

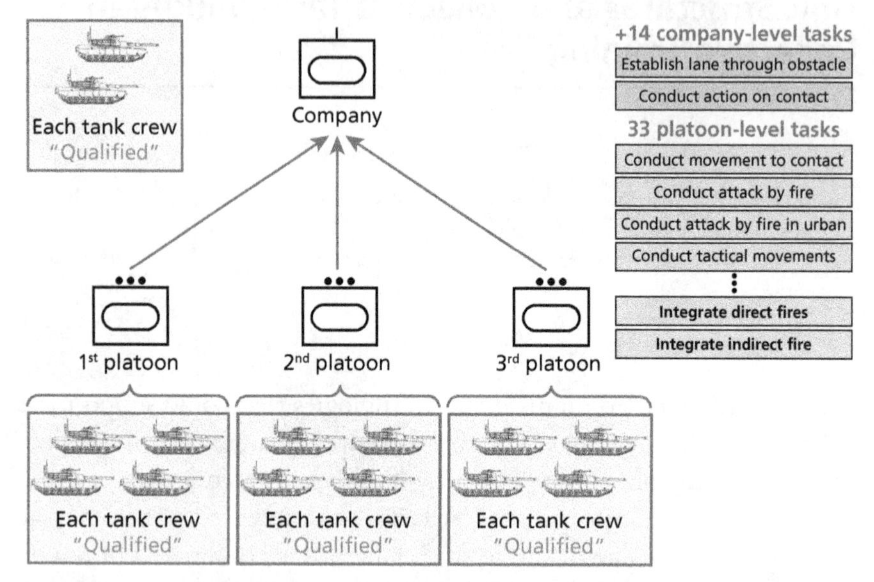

SOURCE: Based on the CATS for Armor: Tank Company/Armored Cavalry Regiment (17477L000), Conduct Tank Platoon Operations–Tank Company/Armored Cavalry Squadron (17-TS-3700), in U.S. Army, Combined Arms Training Strategies (CATS) database, accessed via the Army's Digital Training Management System, undated a.

Table A.1
Combined Arms Training Strategy for Platoons

Collective Task Number	Collective Task Title
05-3-1001	Establish a Lane Through an Obstacle Using Mechanical Techniques
07-2-1090	Conduct a Movement to Contact (Platoon-Company)
07-2-1198	Conduct a Mounted Tactical Road March (Platoon-Company)
07-2-1256	Conduct an Attack by Fire (Platoon-Company)
07-2-1261	Conduct an Attack in an Urban Area (Platoon-Company)
07-2-1324	Conduct Area Security (Platoon-Company)
07-2-1342	Conduct Tactical Movement (Platoon-Company)
07-2-1378	Defend in an Urban Area (Platoon-Company)
07-2-1396	Employ Obstacles (Platoon-Company)
07-2-1450	Secure Routes (Platoon-Company)
7-2-3000	Conduct Support by Fire (Platoon-Company)
7-2-3027	Integrate Direct Fires (Platoon-Company)
7-2-3036	Integrate Indirect Fire Support (Platoon-Company)
7-2-4054	Secure Civilians During Operations (Platoon-Company)
7-2-5027	Conduct Consolidation and Reorganization (Platoon-Company)
7-2-5045	Conduct Negotiations (Platoon-Company)
7-2-5063	Conduct Risk Management (Platoon-Company)
7-2-5081	Conduct Troop-leading Procedures (Platoon-Company)
7-2-6045	Employ Deception Techniques (Platoon-Company)
7-2-6063	Maintain Operations Security (Platoon-Company)
7-2-9001	Conduct an Attack (Platoon-Company)
7-2-9002	Conduct a Bypass (Platoon-Company)
7-2-9003	Conduct an Area Defense (Platoon-Company)
7-2-9004	Conduct a Delay (Platoon-Company)
7-2-9005	Conduct a Linkup (Platoon-Company)
7-2-9006	Conduct a Passage of Lines as the Passing Unit (Platoon-Company)
7-2-9007	Conduct a Passage of Lines as the Stationary Unit (Platoon-Company)

Table A.1—Continued

Collective Task Number	Collective Task Title
7-2-9012	Conduct a Relief in Place (Platoon-Company)
7-2-9014	Occupy an Assembly Area (Platoon-Company)
7-2-9051	Conduct a Cordon and Search (Platoon-Company)
7-3-9013	Conduct Action on Contact
7-3-9016	Establish an Observation Post
7-3-9022	Conduct a Security Patrol
08-2-0003	Treat Casualties
08-2-0004	Evacuate Casualties
17-2-2633	Secure a Basecamp (Platoon-Company)
17-2-3070	Breach an Obstacle (Platoon-Company)
17-2-9225	Conduct a Screen (Platoon-Company)
17-3-3809	Conduct Battle Handover
17-5-5517	Employ a Mine Clearing Blade on an M1-Series Tank
17-5-5518	Employ a Mine Clearing Roller on a M1-Series Tank
19-3-1301	Conduct Dislocated Civilian Control
19-3-2007	Conduct Convoy Security
19-3-2406	Conduct Roadblock and Checkpoint
19-3-3107	Process Detainees at Point of Capture
63-2-4546	Conduct Logistics Package Support
71-2-5311	Integrate Soldier and Leader Engagement into Small Unit Operations

SOURCE: U.S. Army, Combined Arms Training Strategies (CATS) database, accessed via the Army's Digital Training Management System, undated a.

APPENDIX B

Interview and Focus Group Questions

1. How easy/difficult are the following aspects of using CCTT/AVCATT/VBS3 for collective training?
   (For each: What is the process? How well does it work? What are the positive experiences or challenges you have encountered?)
   a. Scheduling equipment/facility
   b. Scheduling/coordinating with training staff, e.g., observer-controllers
   c. Accessing or designing training scenarios
   d. Integrating with other systems, such as Blue Force Tracker
   e. Conducting air-ground operations
   f. Obtaining data about collective performance
   g. Maintenance of hardware/software
   h. Uptime for hardware/software
2. How realistic is training in CCTT/AVCATT/VBS3 for
   a. Creating realistic combat stress, such as sensory overload or disorientation
   b. Replicating equipment (e.g., current Abrams MBT [main battle tank] or BFV) functionality
   c. Replicating equipment form and fit
   d. Replicating communication capabilities
   e. Replicating environmental conditions and effects
   f. Replicating weapons systems and effects
   g. Replicating terrain and effects

3. To what extent does CCTT/AVCATT/VBS3 enable personnel to get sufficient practice conducting collective tasks?
4. How do you think about the trade-off between the collective training experiences for a
   a. Platoon of Armor/BFV in CCTT versus VBS3?
   b. Lead and wingman or other multiple aircraft maneuvers in AVCATT versus VBS3?
   c. The number of repetitions in CCTT/AVCATT versus in VBS3 suites?
   d. The variation of scenarios in CCTT/AVCATT versus in VBS3 suites?
5. How effective is CCTT/AVCATT/VBS3 for training novices? Experienced personnel?
6. How effective is CCTT/AVCATT/VBS3 for preparing personnel for collective tasks during:
   a. Shooting tank tables? Flight training?
   b. Combined arms live fire exercise?
   c. CTC rotations?
   d. Combat?
7. Any additional comments about use of CCTT/AVCATT/VBS3 for collective training?

APPENDIX C

Survey of Training Consumers

Consent to Participate in RAND Survey of Virtual Collective Training Technologies

Purpose. The U.S. Army Training and Doctrine Command, Combined Arms Center, asked the RAND Corporation, a nonprofit research organization, to assist in an evaluation of virtual collective training (VCT) technologies. We are asking you to participate in a survey about your attitudes toward and experience using VCT technologies. The survey takes approximately 15 minutes to complete.

Participation. Your participation in the survey is voluntary. You may refuse to participate, skip questions that you don't want to answer, or stop participating at any time without penalty.

Confidentiality. Your participation in the survey is anonymous. No one at RAND or in the Army can link your identity to your survey responses.

How your answers will be used. Your answers will be combined with other participants' responses. RAND will report only aggregate results to the Army.

Risks and benefits. There are no risks to completing the survey. There are no direct benefits to you for participating, but your involvement will contribute to the Army's understanding of the effectiveness of VCT technologies.

Whom to contact. If you have questions or comments about the evaluation, please contact the project lead: Dr. Susan Straus, 412-683-2300, x4925, sgstraus@rand.org, RAND Corporation, 4570 Fifth Avenue, Pittsburgh, PA 15213. If you have questions or concerns about your rights as a participant in this evaluation, contact Carolyn Tschopik, Human Subjects Protection Committee, RAND, 1700 Main Street, Santa Monica, CA, 90407, 310-393-0411, x6124.

1. If you agree to participate, select "Yes" and click "Next" to begin the survey.

O Yes

O No

2. I am familiar with the Aviation Combined Arms Tactical Trainer (AVCATT).

O Yes

O No

3. Please indicate your familiarity with AVCATT.

Check all that apply.

O I have planned rotary wing aviation collective unit training using AVCATT

O I have participated in rotary wing aviation collective unit training using AVCATT

O I have observed rotary wing aviation collective unit training using AVCATT

O I have heard about AVCATT from other Army personnel

4. You indicated that you have *planned* collective unit training using AVCATT. Approximately, how many such training sessions have you planned?

____

5. You indicated that you have *participated in* collective unit training using AVCATT. Approximately, how many such training sessions have you participated in?

____

6. You indicated that you have planned training or trained in AVCATT. Please select the airframes that you have *trained in or planned training in* using AVCATT.

Check all that apply.

O AH-64D

O AH-64E

O CH-47D

O CH-47F

O OH-58

O UH-60L

O UH-60M

O UH-72

O Other, please specify

____

7. Please indicate the extent to which you disagree or agree with the following statements about *AVCATT for collective* training.

If you do not feel that you can judge a particular item, please select the "Not able to judge" option on the right.

	Strongly Disagree	Disagree	Agree	Strongly Agree	Not Able to Judge
Allows units to conduct many iterations of collective task training in a short period of time	O	O	O	O	O
Enables useful after action reviews (AARs) for collective task training	O	O	O	O	O
Is effective for collective task training in air-ground operations	O	O	O	O	O
Facilitates rapid planning and preparation of collective task training events	O	O	O	O	O
Easily interfaces with the Aviation Mission Planning System (AMPS) for pre-mission planning	O	O	O	O	O
Effectively replicates the functionality (e.g., speed, mobility, weapon accuracy) of a current Army rotary wing aircraft	O	O	O	O	O
Effectively replicates the form and fit (e.g., physical interior) of a current Army rotary wing aircraft	O	O	O	O	O
Integrates well with other training systems for collective task training, such as Blue Force Tracker or unmanned aerial system	O	O	O	O	O
Is effective for collective task training for novice personnel	O	O	O	O	O
Is effective for collective task training for experienced personnel	O	O	O	O	O
Requires a lot of time to coordinate collective training with AVCATT staff	O	O	O	O	O
Allows user to physically manipulate controls with appropriate responses	O	O	O	O	O

	Strongly Disagree	Disagree	Agree	Strongly Agree	Not Able to Judge
Replicates effects of environmental conditions (e.g., rain, snow, wind, dust, night/day) on aircraft, sensor, and weapon system performance	O	O	O	O	O
Allows development of specific terrain for collective task training	O	O	O	O	O
Replicates various weapon and sensor systems and effects (e.g., fire control radar, manned-unmanned teaming, enemy vehicles, jamming, cyber-attack, GPS degrade) for collective task training	O	O	O	O	O
Creates realistic types of combat stress for collective task training, for example sensory or cognitive overload or disorientation	O	O	O	O	O
Provides realistic impacts of environmental conditions on platform performance (e.g., dust, sand, hot & high, fog, smoke) for collective task training	O	O	O	O	O
Provides high fidelity visual scenes to support collective task training	O	O	O	O	O
Provides accurate weapon and sensor system performance for collective task training	O	O	O	O	O
Provides accurate flight model performance for collective task training	O	O	O	O	O
Provides accurate communication capabilities (e.g., networking, digital traffic passage) required for collective task training	O	O	O	O	O
Provides accurate direct and distributed target acquisition performance for collective task training	O	O	O	O	O
Is available for collective task training when needed	O	O	O	O	O
Is easy to set up for collective task training	O	O	O	O	O

	Strongly Disagree	Disagree	Agree	Strongly Agree	Not Able to Judge
Can be used to train a broad variety of collective task training scenarios	O	O	O	O	O
Provides clear feedback/ information (e.g., automated scores) about crews' performance on required collective tasks	O	O	O	O	O
Supports assessment of communication among participants in collective training	O	O	O	O	O
Requires a lot of unit personnel to prepare for collective task training	O	O	O	O	O
Requires a lot of unit personnel to provide and supervise collective task training	O	O	O	O	O
Allows units to get sufficient practice to develop proficiency on key collective tasks	O	O	O	O	O
Is valuable for preparing personnel for collective tasks during Combat Training Center (CTC) rotations	O	O	O	O	O
Is valuable for preparing personnel for collective tasks in combat	O	O	O	O	O
My supervisor thinks use of AVCATT is important for collective training	O	O	O	O	O

8. I am familiar with Virtual Battlespace 3 (VBS3).

O Yes
O No

9. Please indicate your familiarity with VBS3.

Check all that apply.

O I have planned rotary wing aviation collective unit training using VBS3
O I have participated in rotary wing aviation collective unit training using VBS3
O I have observed rotary wing aviation collective unit training using VBS3
O I have heard about VBS3 from other Army personnel

10. You indicated that you have *planned* collective unit training using VBS3. Approximately, how many such training sessions have you planned?

____

11. You indicated that you have *participated in* collective unit training using VBS3. Approximately, how many such training sessions have you participated in?

____

12. You indicated that you have planned training or trained in VBS3. Please select the airframes that you have *trained in or planned training in* using VBS3.

Check all that apply.

O AH-64D
O AH-64E
O CH-47D
O CH-47F
O OH-58
O UH-60L
O UH-60M
O UH-72
O Other, please specify

____

13. Please indicate the extent to which you disagree or agree with the following statements about *VBS3 for collective* training.

If you do not feel that you can judge a particular item, please select the "Not able to judge" option on the right.

	Strongly Disagree	Disagree	Agree	Strongly Agree	Not Able to Judge
Allows units to conduct many iterations of collective task training in a short period of time	O	O	O	O	O
Enables useful after action reviews (AARs) for collective task training	O	O	O	O	O
Is effective for collective task training in air-ground operations	O	O	O	O	O

	Strongly Disagree	Disagree	Agree	Strongly Agree	Not Able to Judge
Facilitates rapid planning and preparation of collective task training events	O	O	O	O	O
Easily interfaces with the Aviation Mission Planning System (AMPS) for pre-mission planning.	O	O	O	O	O
Effectively replicates the functionality (e.g., speed, mobility, weapon accuracy) of a current Army rotary wing aircraft	O	O	O	O	O
Effectively replicates the form and fit (e.g., physical interior) of a current Army rotary wing aircraft	O	O	O	O	O
Integrates well with other training systems for collective task training, such as Blue Force Tracker or unmanned aerial system	O	O	O	O	O
Is effective for collective task training for novice personnel	O	O	O	O	O
Is effective for collective task training for experienced personnel	O	O	O	O	O
Requires a lot of time to coordinate collective training with VBS3 staff	O	O	O	O	O
Allows user to physically manipulate controls with appropriate responses	O	O	O	O	O
Replicates effects of environmental conditions (e.g., rain, snow, wind, dust, night/day) on aircraft, sensor, and weapon system performance	O	O	O	O	O
Allows development of specific terrain for collective task training	O	O	O	O	O
Replicates various weapon and sensor systems and effects (e.g., fire control radar, manned-unmanned teaming, enemy vehicles, jamming, cyber-attack, GPS degrade) for collective task training	O	O	O	O	O

	Strongly Disagree	Disagree	Agree	Strongly Agree	Not Able to Judge
Creates realistic types of combat stress for collective task training, for example sensory or cognitive overload or disorientation	O	O	O	O	O
Provides realistic impacts of environmental conditions on platform performance (e.g., dust, sand, hot & high, fog, smoke) for collective task training	O	O	O	O	O
Provides high fidelity visual scenes to support collective task training	O	O	O	O	O
Provides accurate weapon and sensor system performance for collective task training	O	O	O	O	O
Provides accurate flight model performance for collective task training	O	O	O	O	O
Provides accurate communication capabilities (e.g., networking, digital traffic passage) required for collective task training	O	O	O	O	O
Provides accurate direct and distributed target acquisition performance for training collective task training	O	O	O	O	O
Is available for collective task training when needed	O	O	O	O	O
Is easy to set up for collective task training	O	O	O	O	O
Can be used to train a broad variety of collective task training scenarios	O	O	O	O	O
Provides clear feedback/information (e.g., automated scores) about crews' performance on required collective tasks	O	O	O	O	O
Supports assessment of communication among participants in collective training	O	O	O	O	O
Requires a lot of unit personnel to prepare for collective task training	O	O	O	O	O

	Strongly Disagree	Disagree	Agree	Strongly Agree	Not Able to Judge
Requires a lot of unit personnel to provide and supervise collective task training	O	O	O	O	O
Allows units to get sufficient practice to develop proficiency on key collective tasks	O	O	O	O	O
Is valuable for preparing personnel for collective tasks during Combat Training Center (CTC) rotations	O	O	O	O	O
Is valuable for preparing personnel for collective tasks in combat	O	O	O	O	O
My supervisor thinks use of VBS3 is important for collective training	O	O	O	O	O

14. Have you ever used a commercial simulation game, other than VBS3, for training Army rotary wing aircraft skills?

	Yes	No	No, but I know someone who has
Individual-level skills	O	O	O
Crew-level skills	O	O	O
Collective skills at platoon or company level	O	O	O

15. Please provide any comments you have about the use of commercial simulation games for training.

16. Please indicate your experience with the following methods to train crew and collective tasks. You may check all that apply.

	Personally used for *crew-level* training	Personally used for *PLT-level* collective training	Know fellow soldier who has used for *crew-level* training	Know fellow soldier who has used for *PLT-level* collective training
"Chair Drill" for communication skills	O	O	O	O
Training Aids (e.g., engine cutaway, hydraulic board)	O	O	O	O
Procedural Trainer, (e.g., gunnery trainer, cockpit procedural trainer)	O	O	O	O
Other (indicate below)	O	O	O	O

17. If you selected "other" above, please indicate which method here:

____

18. Please indicate the extent to which you disagree or agree with the following statements about virtual training in general:

	Strongly Disagree	Disagree	Agree	Strongly Agree
Desk-top simulators can enhance the effectiveness of live training for mission-critical collective tasks	O	O	O	O
Physical simulators can enhance the effectiveness of live training for mission-critical collective tasks	O	O	O	O
Simulators require high visual fidelity to effectively train mission-critical collective tasks	O	O	O	O
Simulators require high physical fidelity to effectively train mission-critical collective tasks	O	O	O	O

19. Please indicate if you received training on how to plan virtual collective training events.

Check all that apply.

O Professional Military Education Course such as the Captain's Career Course

O Formal class provided by the Training Support Center or Mission Training Complex

O Informal on-the-job training

O Other, please describe:

____

20. How many hours a week do you play action video games?

Select from the drop-down menu.

21. How many hours a week do you play multi-player competitive video games?

Select from the drop-down menu.

22. I consider myself . . .

O not a gamer

O a casual gamer

O a serious gamer

23. Please answer the following questions about your experience with video games.

For your reference:

MMOG (massively multiplayer online game) - A game supporting large numbers of players that exists in a virtual environment that persists regardless of the user's presence (e.g., MMORPGs like World of Warcraft or Second Life, MMOFPS like Destiny or CrossFire)

MOBA (multiplayer online battle arena) - a game supporting large numbers of players that exists in a virtual environment that only persists for the duration of a battle (e.g., League of Legends, DotA, Call of Duty, Starcraft PvP)

	Yes	No
I play video games that replicate ground maneuver combat	O	O
I play video games that replicate aviation	O	O
I have played in a gaming tournament	O	O
I have built my own computer for gaming	O	O
I play MMOGs	O	O
I play MOBAs	O	O

	Yes	No
I discuss games with friends or with other players on online forums	O	O
I read game-related information through sources such as blogs, magazine, and industry web sites	O	O
I typically modify commercial games rather than playing them as is	O	O
I am an early adopter of technology	O	O
I am willing to pay for games or subscriptions to online games	O	O
I have waited in line to buy a game on the first day it was released	O	O
I typically have the latest gaming platforms	O	O
I am willing to spend a great deal of time to learn to play complicated games	O	O
I have attended one or more gaming conferences or competitions	O	O
My goal in playing games is to beat/finish the game or climb rank in competitive player-versus-player games	O	O

24. What is your rank?

Select from the drop-down menu.

25. Please indicate the position(s) you have held and/or currently hold.

Check all that apply.

O Platoon leader
O Company commander
O Brigade/battalion staff officer
O Other, please specify

____

26. What is your career field?

O Armor
O Cavalry
O Aviation
O Infantry
O Other, please specify

____

27. Please indicate which airframe(s) you operate in your unit.

Check all that apply.

O AH-64D
O AH-64E
O CH-47D
O CH-47F
O OH-58
O UH-60L
O UH-60M
O UH-72
O Other, please specify

____

28. What are your total actual flight hours?

____

29. Please select your current service.

O Active
O Reserve
O National Guard
O Other, please specify

____

30. How many years have you been in the Army?

Select from the drop-down menu.

31. How many combat tours have you served?

O None
O 1
O 2
O 3
O 4+

32. How many Combined Training Center (CTC) rotations have you completed?

O None

O 1

O 2

O 3

O 4+

33. What is your age? (years)

____

34. From my perspective, the way to improve the effectiveness of virtual or computer-based collective aviation training would be to . . .

35. Please provide any additional comments about Army virtual or computer-based collective training, including suggestions for improvement.

Abbreviations

AAR	after-action review
ABCT	armor brigade combat team
AC	active component
AR	Army Regulation
ARNG	Army National Guard
ATOM	Anti-Air Teamwork Observation Measure
AVCATT	Aviation Combined Arms Tactical Trainer
BFV	Bradley Fighting Vehicle
CAC-T	Combined Arms Center for Training
CATS	Combined Arms Training Strategy
CATT	Combined Arms Tactical Trainer
CCTT	Close Combat Tactical Trainer
CLS	contractor logistics support
CONUS	continental United States
COTS	commercial-off-the-shelf
CRM	crew resource management

CTC	combat training center
DSTS	Dismounted Soldier Training System
EBAT	Event-Based Approach to Training
ECG	electrocardiogram
EEG	electroencephalogram
ERT	emergency response team
FY	fiscal year
GAO	Government Accountability Office
GFT	Games for Training
GOTS	government-off-the-shelf
MCCTT	mobile Close Combat Tactical Trainer
MCoE	Maneuver Center of Excellence
MTS	multiteam system
NCO	noncommissioned officer
NTC	National Training Center
Objective-T	Objective Assessment of Training Proficiency
OCONUS	outside the continental United States
PC	personal computer
PEO STRI	Program Executive Office for Simulation, Training and Instrumentation
POM	program objective memorandum
PSME	physically simulated military equipment
RC	reserve component
RDT&E	research, development, test, and evaluation

RVTT	Reconfigurable Vehicle Tactical Trainer
SBT	simulation-based training
SD	standard deviation
SIMNET	simulator networking
SME	subject-matter expert
STE	Synthetic Training Environment
T&EO	training and evaluation outline
TADMUS	Tactical Decision Making Under Stress
TADSS	training aids, devices, simulators, and simulations
TARGET	Targeted Acceptable Responses to Generated Events or Tasks
TCM	TRADOC Capability Manager
TP	TRADOC pamphlet
TRADOC	U.S. Army Training and Doctrine Command
TSAID	Training Support Analysis and Integration Division
TSP	training support package
TTRAM	Task and Training Requirements Analysis Methodology
UK	United Kingdom
USAACE	U.S. Army Aviation Center of Excellence
USAF	U.S. Air Force
V&G	virtual and gaming
VBS3	Virtual Battlespace 3
VME	virtual military equipment

References

Air Force Instruction 11-2F-22A, Vol. 1, *Flying Operations: F-22A—Aircrew Training*, Washington, D.C.: Department of the Air Force, May 19, 2006.

Allen, John A., Robert T. Hays, and Louis C. Buffardi, "Maintenance Training Simulator Fidelity and Individual Differences in Transfer of Training," *Human Factors*, Vol. 28, No. 5, October 1986, pp. 497–509.

Alliger, George M., Scott I. Tannenbaum, Winston Bennett, Holly Traver, and Allison Shotland, "A Meta-Analysis of the Relations Among Training Criteria," *Personnel Psychology*, Vol. 50, No. 2, June 1997, pp. 341–358.

Alvarez, Kaye, Eduardo Salas, and Christina M. Garofano, "An Integrated Model of Training Evaluation and Effectiveness," *Human Resource Development Review*, Vol. 3, No. 4, December 2004, pp. 385–416.

Army Regulation 71-9, *Warfighting Capabilities Determination*, Washington, D.C.: Headquarters, Department of the Army, December 28, 2009.

Army Regulation 350-38, *Policies and Management for Training Aids, Devices, Simulators, and Simulations*, Washington, D.C.: Headquarters, Department of the Army, March 28, 2013.

Army Regulation 350-1, *Army Training and Leader Development*, Washington, D.C.: Headquarters, Department of the Army, August 19, 2014.

Army Regulation 350-52, *Army Training Support System*, Washington, D.C.: Headquarters, Department of the Army, January 17, 2014.

Arthur, Winfred, Jr., Winston Bennett, Jr., Pamela S. Edens, and Suzanne T. Bell, "Effectiveness of Training in Organizations: A Meta-Analysis of Design and Evaluation Features," *Journal of Applied Psychology*, Vol. 88, No. 2, April 2003, pp. 234–245.

Beaubien, Jeffrey M., and David P. Baker, "The Use of Simulation for Training Teamwork Skills in Health Care: How Low Can You Go?" *Quality and Safety in Health Care*, Vol. 13, No. 1, October 2004, pp. i51–i56.

Bell, Bradford S., Scott I. Tannenbaum, J. Kevin Ford, Raymond A. Noe, and Kurt Kraiger, "100 Years of Training and Development Research: What We Know and Where We Should Go," *Journal of Applied Psychology*, Vol. 102, No. 3, March 2017, pp. 305–323.

Benishek, Lauren E., Elizabeth H. Lazzara, William L. Gaught, Lygia L. Arcaro, Yasuharu Okuda, and Eduardo Salas, "The Template of Events for Applied and Critical Healthcare Simulation (TEACH Sim): A Tool for Systematic Simulation Scenario Design," *Simulation in Healthcare*, Vol. 10, No. 1, February 2015, pp. 21–30.

Bohemia Interactive, "Arma," webpage, undated. As of January 25, 2017: https://arma3.com/

Bohemia Interactive Simulations, "Games for Training: U.S. Army," webpage, undated. As of January 25, 2017: https://bisimulations.com/showcase/us-army-gft

———, "Bohemia Interactive Simulations Developing High-Fidelity,Game-Based Image Generator," webpage, September 15, 2014. As of March 17, 2017: https://blogit.realwire.com/Bohemia-Interactive-Simulations-Developing-High-Fidelity-Game-Based-Image

Bowers, Clint A., and Florian Jentsch, "Use of Commercial, Off-the-Shelf, Simulations for Team Research," *Advances in Human Performance and Cognitive Engineering Research*, Vol. 1, May 2001, pp. 293–318.

Bowers, Clint, Eduardo Salas, Carolyn Prince, and Michael Brannick, "Games Teams Play: A Method for Investigating Team Coordination and Performance," *Behavior Research Methods, Instruments, & Computers*, Vol. 24, No. 4, December 1992, pp. 503–506.

Brown, Bill, Stephen Wilkinson, John Nordyke, David Riede, and Steve Huysson, *Developing an Automated Training Analysis and Feedback System for Tank Platoons*, Alexandria, Va.: U.S. Army Research Institute for the Behavioral and Social Sciences, Research Report 1708, May 1997.

Burnside, Billy L., *Assessing the Capabilities of Training Simulations: A Method and Simulation Networking (SIMNET) Application*, Alexandria, Va.: Army Research Institute for the Behavioral and Social Sciences, Research Report 1565, June 1990.

Bymer, Loren, "DSTS: First Immersive Virtual Training System Fielded," webpage, August 1, 2012. As of January 25, 2017: https://www.army.mil/article/84728/DSTS

CAC-T—*See* Combined Arms Center–Training.

Cannon-Bowers, Janis A., and Clint Bowers, "Synthetic Learning Environments: On Developing a Science of Simulation, Games, and Virtual Worlds for Training," in Steve W. J. Kozlowski and Eduardo Salas, eds., *Learning, Training, and Development in Organizations*, New York: Taylor & Francis Group, 2009, pp. 229–261.

Cannon-Bowers, Janis A., and Eduardo Salas, "Individual and Team Decision Making Under Stress: Theoretical Underpinnings," in Janis A. Cannon-Bowers and Eduardo Salas, eds., *Making Decisions Under Stress: Implications for Individual and Team Training*, Washington, D.C.: American Psychological Association, 1998, pp. 17–38.

Cascio, Wayne F., "Using Utility Analysis to Assess Training Outcomes," in Irwin L. Goldstein, ed., *Training and Development Organizations*, San Francisco: Jossey-Bass, 1989, pp. 63–88.

Champney, Roberto K., Meredith Carroll, and Glenn Surpris, "A Human Experience Approach to Optimizing Simulator Fidelity," *Proceedings of the Human Factors and Ergonomics Society 58th Annual Meeting*, Chicago: Human Factors and Ergonomics Society, October 2014, pp. 2355–2359.

Chapman, Robert, "Accreditation Policy and Practice for Immersive Warfighter Simulators," paper presented at the Interservice/Industry Simulation Education and Training Conference, Orlando, December 4–7, 2006.

Chapman, Robert, and Charles Colegrove, "Transforming Operational Training in the Combat Air Forces," *Military Psychology*, Vol. 25, No. 3, 2013, pp. 177–190.

Chen, David Yu, Randy Jensen, Oscar Bascara, and Nancy Harmon, "Enabling Automated AAR Development by Abstracting Data Collection from Analysis," paper presented at the Interservice/Industry Training, Simulation, and Education Conference, 2007. As of January 19, 2017:
http://citeseerx.ist.psu.edu/viewdoc/summary?doi=10.1.1.587.3322

Cohen, Iris, Willem-Paul Brinkman, and Mark A. Neerinex, "Modelling Environmental and Cognitive Factors to Predict Performance in a Stressful Training Scenario on a Naval Ship Simulator," *Cognition, Technology, and Work*, Vol. 17, No. 4, November 2015, pp. 503–519.

Combined Arms Center–Training, *Close Close Combat Tactical Trainer and Games for Training Cost-Benefit Analysis*, Fort Eustis, Va.: Army Training Support Center, Training Support Analysis and Integration Directorate, November 18, 2014.

"Call of Duty Tracker," website, last updated January 24, 2016. As of August 22, 2018:
http://codtracker.net/aw/player/xbl/TheKingDezire

De Vries, Thomas A., John R. Hollenbeck, Robert B. Davison, Frank Walter, and Gerben S. van der Vegt, "Managing Coordination in Multiteam Systems: Integrating Micro and Macro Perspectives," *Academy of Management Journal*, Vol. 59, No. 5, 2016, pp. 1823–1844.

De Winter, Joost C. F., Dimitra Dodou, and Max Mulder, "Training Effectiveness of Whole Body Flight Simulator Motion: A Comprehensive Meta-Analysis," *International Journal of Aviation Psychology*, Vol. 22, No. 2, 2012, pp. 164–183.

Dietz, Aaron S., Wendy L. Bedwell, James M. Oglesby, Eduardo Salas, and Kathryn E. Keeton, "Synthetic Task Environments for Improving Performance at Work: Principles and the Road Ahead," in Jose M. Cortina and Ronald S. Landis, eds., *Modern Research Methods for the Study of Behavior in Organizations*, New York: Routledge, 2013, pp. 349–380.

DoD—*See* U.S. Department of Defense.

Doughty, Philip L., Hervey W. Stern, and Cindy Thompson, *Guidelines for Cost-Effectiveness Analysis for Navy Training and Education*, San Diego: Navy Personnel Research and Development Center, NPRDC-SR-76TQ-12, 1976.

Dreyfus, Hubert, and Stuart E. Dreyfus, *Mind over Machines*, New York: Free Press, 1986.

Driskell, James E., Joan H. Johnston, and Eduardo Salas, "Does Stress Training Generalize to Novel Settings?" *Human Factors*, Vol. 43, No. 1, Spring 2001, pp. 99–110.

Dwyer, Daniel J., Joan K. Hall, Catherine Volpe, Janis A. Cannon-Bowers, and Eduardo Salas, *A Performance Assessment Task for Examining Tactical Decision Making Under Stress*, Orlando: Naval Training Systems Center, NAVTRASYSCEN-SR-92-002, 1992.

Dwyer, Daniel J., Randall L. Oser, Eduardo Salas, and Jennifer E. Fowlkes, "Performance Measurement in Distributed Environments: Initial Results and Implications for Training," *Military Psychology*, Vol. 11, No. 2, 1999, pp. 189–215.

Dwyer, Daniel J., and Eduardo Salas, "Principles of Performance Measurement for Ensuring Aircrew Training Effectiveness," in Harold F. O'Neil and Dee H. Andrews, eds., *Aircrew Training and Assessment*, Mahwah, N.J.: Lawrence Erlbaum Associates, 2000, pp. 223–244.

Endsley, Mica R., "Theoretical Underpinnings of Situation Awareness: A Critical Review," in Mica R. Endsley and Daniel J. Garland, eds., *Situation Awareness Analysis and Measurement*, Mahwah, N.J.: Lawrence Erlbaum Associates, 2000, pp. 3–32.

Entin, Elliot E., and Daniel Serfaty, "Adaptive Team Coordination," *Human Factors*, Vol. 41, No. 2, 1999, pp. 312–325.

Ericsson, K. Anders, "The Influence of Experience and Deliberate Practice on the Development of Superior Expert Performance," in K. Anders Ericsson, Neil Charness, Paul J. Feltovich, and Robert R. Hoffman, eds., *The Cambridge Handbook of Expertise and Expert Performance*, New York: Cambridge University Press, 2006, pp. 683–703.

Ericsson, K. Anders, Ralf T. Krampe, and Clemens Tesch-Römer, "The Role of Deliberate Practice in the Acquisition of Expert Performance," *Psychological Review*, Vol. 100, No. 3, July 1993, pp. 363–406.

Field Manual 17-18, *Light Armor Operations*, Washington, D.C.: Headquarters, Department of the Army, 1994.

Ford, J. Kevin, and Aaron M. Schmidt, "Emergency Response Training: Strategies for Enhancing Real-World Performance," *Journal of Hazardous Materials*, Vol. 75, No. 2, June 2000, pp. 195–215.

Fowlkes, Jennifer, Daniel J. Dwyer, Randall L. Oser, and Eduardo Salas, "Event-Based Approach to Training (EBAT)," *International Journal of Aviation Psychology*, Vol. 8, No. 3, July 1998, pp. 209–221.

Fowlkes, Jennifer E., Norman E. Lane, Eduardo Salas, Thomas Franz, and Randall Oser, "Improving the Measurement of Team Performance: The TARGETs Methodology," *Military Psychology*, Vol. 6, No. 1, 1994, pp. 47–61.

Fowlkes, Jennifer E., Norman E. Lane, Eduardo Salas, Randall L. Oser, and Carolyn Prince, "Targets for Aircrew Coordination Training," *14th Interservice/Industry Training Systems and Education Conference*, San Antonio, Tex.: National Training and Simulation Association, 1992, pp. 342–352.

Fowlkes, Jennifer, Jerry Owens, Corbin Hughes, Joan H. Johnston, Michael Stiso, Amanda Hafich, and Kevin Bracken, "Constraint-Directed Performance Measurement for Large Tactical Teams," *Proceedings of the Human Factors and Ergonomics Society 49th Annual Meeting*, Orlando: Human Factors and Ergonomics Society, September 26–30, 2005, pp. 2125–2129.

Galán, Federico Cirett, and Carole R. Beal, "EEG Estimates of Engagement and Cognitive Workload Predict Math Problem Solving Outcomes," *International Conference on User Modeling, Adaptation, and Personalization*, Montreal: User Modeling, July 2012, pp. 51–62.

GAO—*See* Government Accountability Office.

Gately, Michael T., Sharon M. Watts, John W. Jaxtheimer, and Robert J. Pleban, *Dismounted Infantry Decision Skills Assessment in the Virtual Training Environment*, Arlington, Va.: U.S. Research Institute for the Behavioral and Social Science, Technical Report 1155, 2005.

Goldberg, Isadore, and Nidhi Khattri, *A Review of Models of Cost and Training Effectiveness Analysis (CTEA)*, Vol. 1: *Training Effectiveness Analysis*, Washington, D.C.: U.S. Army Research Institute for the Behavioral and Social Sciences, ARI Research Note 87-58, 1987.

Gorski, Bruce, and Brian Parrish, *Military Equipment Framework: Synthetic Training Environment*, Leavenworth, Kan.: MITRE, MP160204, 2017.

Gourley, Scott R., “Stimulating Simulation: Technology Advances and Upgrades Boost Realism in Soldier Training,” Association of the United States Army website, February 16, 2016. As of October 25, 2018:
https://www.ausa.org/articles/stimulating-simulation-technology-advances-and-upgrades-boost-realism-soldier-training

Government Accountability Office, *Better Performance and Cost Data Needed to More Fully Assess Simulation-Based Efforts*, Washington, D.C., GAO-13-698, 2013.

———, *Efforts to Adjust Training Requirements Should Consider the Use of Virtual Training Devices*, Washington, D.C., GAO-16-636, 2016.

Graafland, Maurits, Jan M. Schraagen, and Marlies P. Schijven, “Systematic Review of Serious Games for Medical Education and Surgical Skills Training,” *British Journal of Surgery*, Vol. 99, No. 10, 2012, pp. 1322–1330.

Guru, Khurshid A., Ehsan T. Esfahani, Syed J. Raza, Rohit Bhat, Katy Wang, Yana Hammond, Gregory Wilding, James O. Peabody, and Ashirwad J. Chowriappa, “Cognitive Skills Assessment During Robot-Assisted Surgery: Separating the Wheat from the Chaff,” *BJU International*, Vol. 115, No. 1, 2015, pp. 166–174.

Hallmark, Bryan W., and S. Jamie Gayton, *Improving Soldier and Unit Effectiveness with the Stryker Brigade Combat Team Warfighters' Forum*, Santa Monica, Calif.: RAND Corporation, TR-919-A, 2011. As of February 6, 2017:
http://www.rand.org/pubs/technical_reports/TR919.html

Hamstra, Stanley J., Ryan Brydges, Rose Hatala, Benjamin Zendejas, and David A. Cook, “Reconsidering Fidelity in Simulation-Based Training,” *Academic Medicine*, Vol. 89, No. 3, March 2014, pp. 387–392.

Hays, Robert T., John W. Jacobs, Carolyn Prince, and Eduardo Salas, “Flight Simulator Training Effectiveness: A Meta-Analysis,” *Military Psychology*, Vol. 4, No. 2, 1992, pp. 63–74.

Hays, Robert T., and Michael J. Singer, *Simulation Fidelity in Training System Design: Bridging the Gap Between Reality and Training*, New York: Springer-Verlag, 1989.

Headquarters, Department of the Army, *Department of Defense Fiscal Year (FY) 2016 President's Budget Submission: Research, Development, Test & Evaluation, Army RDT&E*, Vol. II: *Budget Activity 5a*, Washington, D.C., February 2015. As of April 17, 2017:
https://www.asafm.army.mil/documents/BudgetMaterial/fy2016/vol5a.pdf

———, *Leader's Guide to Objective Assessment of Training Proficiency (Initial Operating Capability (IOC))*, Washington, D.C.: 2017.

IFC International, *Optimizing the Mix Between Virtual and Live Military Training*, Fairfax, Va., 2013.

Johnson, Robin R., Chris Berka, David Waldman, Pierre Balthazard, Nicola Pless, and Thomas Maak, "Neurophysiological Predictors of Team Performance," *International Conference on Augmented Cognition*, Berlin: Springer, 2013, pp. 153–161.

Johnson, William R., Thomas W. Mastaglio, and Paul D. Peterson, "The Close Combat Tactical Trainer Program," *Proceedings of the 25th Conference on Winter Simulation*, New York: Association for Computing Machinery, 1993, pp. 1021–1029.

Johnston, Joan Hall, Kimberly A. Smith-Jentsch, and Janis A. Cannon-Bowers, "Performance Measurement Tools for Enhancing Team Decision Making," in Michael T. Brannick, Eduardo Salas, and Carolyn Prince, eds., *Team Performance Assessment and Measurement: Theory, Methods, and Applications*, New York: Psychology Press, 1997, pp. 311–327.

Johnston, J. H., P. Gamble, D. Patton, S. Fitzhugh, L. Townsend, L. Milham, D. Riddle, et al., "Squad Overmatch for Tactical Combat Casualty Care: Phase II Initial Findings Report," Orlando: Program Executive Office Simulation, Training and Instrumentation, 2016.

Kirkpatrick, Donald L., *Evaluating Training Programs: The Four Levels*, San Francisco: Berrett-Koehler, 1994.

Klyde, David H., Amanda K. Lampton, P. Chase Schulze, Daniel Alvarez, Robin Johnson, and Leah Rowe, "The Real-Flight Approach to Assess Flight Simulator Force Cueing Fidelity," *AIAA Atmospheric Flight Mechanics (AFM) Conference*, Boston: American Institute of Aeronautics and Astronautics, 2013.

Kozlowski, Steve W. J., and Richard P. DeShon, "A Psychological Fidelity Approach to Simulation-Based Training: Theory, Research and Principles," in Samuel G. Schiflett, Linda R. Elliott, Eduardo Salas, and Michael D. Coovert, eds., *Scaled Worlds: Development, Validation, and Applications*, Burlington, Vt.: Ashgate Publishing, 2004, pp. 75–99.

Kraiger, Kurt, J., Kevin Ford, and Eduardo Salas, "Application of Cognitive, Skill-Based, and Affective Theories of Learning Outcomes to New Methods of Training Evaluation," *Journal of Applied Psychology*, Vol. 78, No. 2, 1993, pp. 311–328.

Krusmark, Michael, Brian T. Schreiber, and Winston Bennett, Jr., *The Effectiveness of a Traditional Gradesheet for Measuring Air Combat Team Performance in Simulated Distributed Mission Operations*, Mesa, Ariz.: Air Force Research Laboratory, TR-2004-0090, 2004.

Lawson, Ben D., Amanda M. Kelley, and Jeremy R. Athy, *A Review of Computerized Team Performance Measures to Identify Military-Relevant, Low-to-Medium Fidelity Tests of Small Group Effectiveness During Shared Information Processing*, Fort Rucker, Ala.: U.S. Army Aeromedical Research Laboratory, Report No. 2012-11, 2012.

Liu, Dahai, Nikolas D. Macchiarella, and Dennis A. Vincenzi, "Simulation Fidelity," in Dennis A. Vincenzi, John A. Wise, Mustapha Moulaoua, and Peter A. Hancock, eds., *Human Factors in Simulation and Training*, New York: CRC Press, 2008, pp. 61–73.

MacMillan, Jean, Eileen B. Entin, Rebecca Morley, and Winston Bennett, Jr., "Measuring Team Performance in Complex and Dynamic Military Environments: The SPOTLITE Method," *Military Psychology*, Vol. 25, No. 3, May 2013, pp. 266–279.

Marks, Michelle A., Leslie A. DeChurch, John E. Mathieu, Frederick J. Panzer, and Alexander Alonso, "Teamwork in Multiteam Systems." *Journal of Applied Psychology*, Vol. 90, No. 5, 2005, pp. 964–971.

Martin, Elizabeth L., and Wayne L. Waag, *Contributions of Platform Motion to Simulator Training Effectiveness: Study 1, Basic Contact*, Brooks Air Force Base, Tex.: Air Force Human Resources Laboratory, AFHRL-TR-78-15, 1978a.

———, *Contributions of Platform Motion to Simulator Training Effectiveness: Study 2, Aerobatics*, Brooks Air Force Base, Tex.: Air Force Human Resources Laboratory, AFHRL-TR-78-52, 1978b.

Mastiglio, Thomas, Jeffrey Wilkinson, Phillip N. Jones, James P. Bliss, and John S. Barnett, *Current Practice and Theoretical Foundations of the After Action Review*, Portsmouth, Va.: U.S. Army Research Institute for the Behavioral and Social Sciences, Technical Report 1290, June 2011.

Mathieu, J. E., Michelle A. Marks, and Stephen J. Zaccaro, "Multi-Team Systems," in Neil Anderson, Deniz Ones, Handan Kepir Sinangil, and Chockalingam Viswesvaran, eds., *International Handbook of Work and Organizational Psychology*, London: Sage, 2001, pp. 289–313.

McRobert, Allistair Paul, Joe Causer, John Vassiliadis, Leonie Watterson, James Kwan, and Mark A. Williams, "Contextual Information Influences Diagnosis Accuracy and Decision Making in Simulated Emergency Medicine Emergencies," *BMJ Quality & Safety*, Vol. 22, No. 6, June 2013, pp. 478–484.

Meredith, Lisa S., Carra S. Sims, Benjamin Saul Batorsky, Adeyemi Okunogbe, Brittany L. Bannon, and Craig A. Myatt, *Identifying Promising Approaches to U.S. Army Institutional Change: A Review of the Literature on Organizational Culture and Climate*, Santa Monica, Calif.: RAND Corporation, RR-1588-A, 2017. As of August 15, 2018:
https://www.rand.org/pubs/research_reports/RR1588.html

Milgram, Paul, and Fumio Kishino, "A Taxonomy of Mixed Reality Visual Displays," *IEICE Transactions on Information and Systems*, Vol. 77, No. 12, 1994, pp. 1321–1329.

Nash, Trevor, "UK MoD to Streamline CATT Support," Shephard Press, webpage, April 26, 2016. As of April 17, 2017:
https://www.shephardmedia.com/news/training-simulation/uk-mod-streamline-catt-support/

Noble, Cliff, "The Relationship Between Fidelity and Learning in Aviation Training and Assessment," *Journal of Air Transportation*, Vol. 7, No. 3, 2002, pp. 33–54.

Noble, John L., and Douglas R. Johnson, *Close Combat Tactical Trainer (CCTT) Cost and Training Effectiveness Analysis (CTEA)*, White Sands Missile Range, N.M.: Department of the Army, May 1991. As of March 20, 2017:
http://www.dtic.mil/dtic/tr/fulltext/u2/b173567.pdf

Norman, Geoff, Kelly Dore, and Lawrence Grierson, "The Minimal Relationship Between Simulation Fidelity and Transfer of Learning," *Medical Education*, Vol. 46, No. 7, July 2012, pp. 636–647.

Ormerod, J. C., "British Military Mission Kuwait Question Relative Benefit of Virtual Reality Training," unpublished manuscript, SOI SIM HQ Army Training Division, Simulation Branch, July 2015.

Oser, Randall L., James W. Gualtieri, Janis A. Cannon-Bowers, and Eduardo Salas, "Training Team Problem Solving Skills: An Event-Based Approach," *Computers in Human Behavior*, Vol. 15, No. 3, May 1999, pp. 441–462.

Owen, Christine, Christopher Bearman, Benjamin Brooks, Janine Chapman, Douglas Paton, and Liaquat Hossain, "Developing a Research Framework for Complex Multi-Team Coordination in Emergency Management," *International Journal of Emergency Management*, Vol. 9, No. 1, 2013, pp. 1–17.

PEO STRI—*See* Program Executive Office for Simulation, Training and Instrumentation.

Program Executive Office for Simulation, Training and Instrumentation, "Synthetic Training Environment (STE)," webpage, undated. As of March 20, 2017:
http://www.peostri.army.mil/synthetic-training-environment-ste-

———, *2016 PEO STRI Weapon Systems Handbook*, Orlando, 2016.

Portrey, Antoinette M., Loren B. Keck, and Brian T. Schreiber, *Challenges in Developing a Performance Measurement System for the Global Virtual Environment*, Mesa, Ariz.: Air Force Research Laboratory, TR-2006-0022, 2006.

Prince, Carolyn, and Florian Jentsch, "Aviation Crew Resource Management Training with Low-Fidelity Devices," in Eduardo Salas, Clint A. Bowers, and Eleana Edens, eds., *Improving Teamwork in Organizations: Applications of Resource Management Training*, Mahwah, N.J.: Lawrence Erlbaum Associates, March 2001, pp. 147–164.

Rosen, Michael A., Eduardo Salas, Scott I. Tannenbaum, Peter Provonost, and Heidi B. King, "Simulation-Based Training for Teams in Health Care: Designing Scenarios, Measuring Performance, and Providing Feedback," in Pascale Carayon, ed., *Handbook of Human Factors and Ergonomics in Health Care and Patient Safety*, London: CRC Press, 2012, pp. 573–594.

Sadagic, Amela, Mathias Kölsch, Greg Welch, Chumki Basu, Chris Darken, Juan P. Wachs, Henry Fuchs, Herman Towles, Neil Rowe, and Jan-Michael Frahm, "Smart Instrumented Training Ranges: Bringing Automated System Solutions to Support Critical Domain Needs," *Journal of Defense Modeling and Simulation*, Vol. 10, No. 3, July 2013, pp. 327–342.

Salas, Eduardo, Clint A. Bowers, and Lori Rhodenizer, "It Is Not How Much You Have but How You Use It: Toward a Rational Use of Simulation to Support Aviation Training," *International Journal of Aviation Psychology*, Vol. 8, No. 3, 1998, pp. 197–208.

Salas, Eduardo, and Janis A. Cannon-Bowers, "The Science of Training: A Decade of Progress," *Annual Review of Psychology*, Vol. 52, No. 1, 2001, pp. 471–499.

Salas, Eduardo, John T. Paige, and Michael A. Rosen, "Creating New Realities in Healthcare: The Status of Simulation-Based Training as a Patient Safety Improvement Strategy," *BMJ Quality & Safety*, Vol. 22, No. 6, 2013, pp. 449–452.

Salas, Eduardo, Michael A. Rosen, Janet D. Held, and Johnny J. Weissmuller, "Performance Measurement in Simulation-Based Training a Review and Best Practices," *Simulation & Gaming*, Vol. 40, No. 3, 2009, pp. 328–376.

Schmidt, Frank L., John E. Hunter, and Kenneth Pearlman, "Assessing the Economic Impact of Personnel Programs on Workforce Productivity," *Personnel Psychology*, Vol. 35, No. 2, 1982, pp. 333–347.

Schnell, Thomas, Rich Cornwall, Melissa Walwanis, and Jeff Grubb, "The Quality of Training Effectiveness Assessment (QTEA) Tool Applied to the Naval Aviation Training Context," *International Conference on Foundations of Augmented Cognition*, San Diego: HCI International, 2009, pp. 640–649.

Schnell, Thomas, Nancy Hamel, Alex Postnikov, Jaclyn Hoke, and Angus L. M. Thom McLean III, "Physiological Based Simulator Fidelity Design Guidance," in Thomas E. Pinelli and Leanna Bullock, eds., *Selected Papers Presented at MODSIM World 2011 Conference and Expo*, Hanover, Md.: NASA Center for AeroSpace Information, March 2012. As of January 18, 2017: https://ntrs.nasa.gov/archive/nasa/casi.ntrs.nasa.gov/20130008649.pdf

Schnell, Thomas, Alex Postnikov, and Nancy Hamel, "Neuroergonomic Assessment of Simulator Fidelity in an Aviation Centric Live Virtual Constructive (LVC) Application," *International Conference on Foundations of Augmented Cognition*, Berlin: Springer Heidelberg, 2011, pp. 221–230.

Sciarini, Lee W., Jefferson D. Grubb, and Philip G. Fatolitis, "Cognitive State Assessment: Examination of EEG-Based Measures on a Stroop Task," *Proceedings of the Human Factors and Ergonomics Society Annual Meeting*, Chicago: Human Factors and Ergonomics Society, 2014, pp. 215–219.

Seibert, Melinda K., Frederick J. Diedrich, Jeanine M. Ayers, Courtney Dean, Troy Zeidman, Martin L. Bink, and John E. Stewart, *Addressing Army Aviation Collective Training Challenges with Simulators and Simulation Capabilities*, Arlington, Va.: U.S. Army Research Institute for the Behavioral and Social Sciences, Research Report 1958, 2012.

Smith-Jentsch, Kimberly A., Joan H. Johnston, and Stephanie C. Payne, "Measuring Team-Related Expertise in Complex Environments," in Janis A. Cannon-Bowers and Eduardo Salas, eds., *Making Decisions Under Stress: Implications for Individual and Team Training*, Washington, D.C.: American Psychological Association, 1998, pp. 61–87.

Staller, Mario S., and Benjamin Zaiser, "Developing Problem Solvers: New Perspectives on Pedagogical Practices in Police Use of Force Training," *Journal of Law Enforcement*, Vol. 4, No. 3, February 2015.

Stevens, Ronald, Trysha Galloway, Peter Wang, Chris Berka, Veasna Tan, Thomas Wohlgemuth, Jerry Lamb, and Robert Buckles, "Modeling the Neurodynamic Complexity of Submarine Navigation Teams," *Computational and Mathematical Organization Theory*, Vol. 19, No. 3, 2013, pp. 346–369.

Stewart, John E., David M. Johnson, and William R. Howse, *Fidelity Requirements for Army Aviation Training Devices: Issues and Answers*, Arlington, Va.: U.S. Army Research Institute for the Behavioral and Social Sciences, Research Report 1887, 2008.

Sticha, Paul, Roy Campbell, and C. Mazie Knerr, "Individual and Collective Training in Live, Virtual and Constructive Environments. Training Concepts For Virtual Environments," Alexandria, Va.: U.S. Army Research Institute for the Behavioral and Social Sciences, Study Report 2002-05, 2002.

Stikic, Maja, Chris Berka, Daniel J. Levendowski, Roberto F. Rubio, Veasna Tan, Stephanie Korszen, Douglas Barba, and David Wurzer, "Modeling Temporal Sequences of Cognitive State Changes Based on a Combination of EEG-Engagement, EEG-Workload, and Heart Rate Metrics," *Frontiers in Neuroscience*, Vol. 8, No. 342, November 2014.

Stout, Renee J., Eduardo Salas, Danielle C. Merket, and Clint A. Bowers, "Low-Cost Simulation and Military Aviation Team Training," *American Institute of Aeronautics and Astronautics Modeling and Simulation Technologies Conference and Exhibit*, Boston: American Institute of Aeronautics and Astronautics, 1998, pp. 311–318.

Sung, Dan, "50 Wearable Tech Gamechangers for 2016," webpage, December 14, 2015. As of January 19, 2017:
https://www.wareable.com/wareable50/best-wearable-tech

Swezey, Robert W., Jerry M. Owens, Maureen L. Bergondy, and Eduardo Salas, "Task and Training Requirements Analysis Methodology (TTRAM): An Analytic Methodology for Identifying Potential Training Uses of Simulator Networks in Teamwork-Intensive Task Environments," *Ergonomics*, Vol. 41, No. 11, 1998, pp. 1678–1697.

Thorndike, Edward Lee, *Educational Psychology*, New York: Lemke and Buechner, 1906.

Toups, Zachary O., Andruid Kerne, and William A. Hamilton, "The Team Coordination Game: Zero-Fidelity Simulation Abstracted from Fire Emergency Response Practice," *ACM Transactions on Computer-Human Interaction (ToCHI)*, Vol. 18, No. 4, 2011, pp. 1–37.

TRADOC Pamphlet 350-70-1, *Training Development in Support of the Operational Domain*, Fort Eustis, Va.: U.S. Army Training and Doctrine Command, February 24, 2012.

TRADOC Pamphlet 350-70-13, *System Training Integration*, Fort Eustis, Va.: U.S. Army Training and Doctrine Command, October 27, 2014.

TRADOC Pamphlet 525-8-2, *The U.S. Army Learning Concept for Training and Education 2020-2040*, Fort Eustis, Va.: U.S. Army Training and Doctrine Command, April 2017.

TRADOC Regulation 350-70, *Army Learning Policy and Systems*, Fort Eustis, Va.: U.S. Army Training and Doctrine Command, July 10, 2017.

Tsifetakis, Emmanuel, and Tom Kontogiannis, "Evaluating Non-Technical Skills and Mission Essential Competencies of Pilots in Military Aviation Environments," *Ergonomics*, May 2017, pp. 1–15.

U.S. Army, Combined Arms Training Strategies (CATS) database, accessed via the Army's Digital Training Management System, undated a. As of October 2, 2018:
https://dtms.army.mil/

———, "Reconfigurable Vehicle Tactical Trainer HMMWV/HEMTT (RVTT)," Fort Leonard Wood, Mo., undated b. As of January 25, 2017:
http://www.wood.army.mil/newweb/garrison/Training%20Aids/RVTT.pdf

U.S. Army, "Virtual Battlespace 3," Stand-To website, May 19, 2014. As of September 28, 2018:
https://www.army.mil/standto/archive_2014-05-19

U.S. Army Acquisition Support Center, "Close Combat Tactical Trainer (CCTT)," webpage, undated a. As of March 20, 2017:
http://asc.army.mil/web/portfolio-item/close-combat-tactical-trainer-cctt/

———, "Aviation Combined Arms Tactical Trainer (AVCATT)," webpage, undated b. As of April 17, 2017:
http://asc.army.mil/web/portfolio-item/
peo-stri-aviation-combined-arms-tactical-trainer-avcatt/

U.S. Department of Defense, *Department of Defense Modeling and Simulation (M&S) Glossary*, Alexandria, Va.: Modeling and Simulation Coordination Office, October 1, 2011. As of January 17, 2017:
http://www.acqnotes.com/Attachments/DoD%20M&S%20Glossary%201%20
Oct%2011.pdf

USC Institute for Creative Technologies, "Early Synthetic Prototyping (ESP)," webpage, undated. As of March 17, 2017:
http://ict.usc.edu/prototypes/early-synthetic-prototyping-esp/

Warr, Peter, Catriona Allan, and Kamal Birdi, "Predicting Three Levels of Training Outcome," *Journal of Occupational and Organizational Psychology*, Vol. 72, No. 3, 1999, pp. 351–375.

CPSIA information can be obtained
at www.ICGtesting.com
Printed in the USA
LVHW041659231019
635121LV00014B/214/P